FACILITY MANAGEMENT RISKS AND OPPORTUNITIES

Bev Nutt

Professor of Facility and Environmental Management
University College, London

and

Peter McLennan

Senior Research Fellow and Lecturer
Bartlett School of Graduate Studies
University College, London

Blackwell
Science

DISTRIBUTORS

Marston Book Services Ltd
PO Box 269
Abingdon
Oxon OX14 4YN
(*Orders:* Tel: 01235 465500
         Fax: 01235 465555)

USA
Blackwell Science, Inc.
Commerce Place
350 Main Street
Malden, MA 02148 5018
(*Orders:* Tel: 800 759 6102
              781 388 8250
         Fax: 781 388 8255)

Canada
Login Brothers Book Company
324 Saulteaux Crescent
Winnipeg, Manitoba R3J 3T2
(*Orders:* Tel: 204 224-4068)

Australia
Blackwell Science Pty Ltd
54 University Street
Carlton, Victoria 3053
(*Orders:* Tel: 03 9347 0300
         Fax: 03 9347 5001)

A catalogue record for this title
is available from the British Library

ISBN 0-632-05797-1

For further information on
Blackwell Science, visit our website:
www.blackwell-science.com

Contents

Preface

The year 2000 has provided a unique threshold for all to consolidate the successes of the past and to focus on the opportunities for the future. For young areas of enterprise such as facility management, this has been a particularly significant time. With 'Threshold 2000' in mind, an international Conference "Futures in Property and Facility Management: creating the platform for innovation", was held at University College London in June 1999. This Conference addressed the fundamental challenges and strategic directions for facility management world-wide, bringing together an international audience of leading facility professionals, property clients and academics.

This book provides a comprehensive account of the ideas and insights, topical issues, expert opinions, new directions and concerns that were debated at this international event. It summarises both the consensus and divergent views that emerged. The Conference focusing on four crucial themes:-

- **New Strategic Directions**: exploring the changing priorities, potential scope, future functions and impact of facility management, in providing strategic support to serve the dynamic objectives and needs of business and public organisations, into the 21st century.
- **Future Performance Imperatives**: developing the basis for the next generation of property and facility performance criteria, new management methods, operational procedures and decision techniques with which to face and manage the unprecedented pace of change in business organisation, working practices and technology.
- **Policy and Investment Development**: investigating the key property and facility management issues for a sustainable future and the development of radically new approaches to investment and risk, with expert speculation on alternative policy positions by sector.
- **Promoting Knowledge Exchange**: securing the framework for the exchange of facility management expertise, research results and data, ideas and concepts, methods and techniques, between those in practice, consultancy, education and research, in order to build an international knowledge base to support facility management futures.

In his opening address to the Conference, Mike Jeffries, the Chief Executive of WS Atkins plc, set out the key links that will be crucial to the future development of property and facility management (Jeffries, 1999).

Linking the facilities management industry with market demands:

"There is little doubt that we in the facilities management industry operate in times of great change with traditional methods of operation being regularly challenged and the market demanding innovative solutions to the provision, financing, operation and maintenance of the physical infrastructure which makes up the built environment".

Linking facility management with business performance:

"What can be stated with confidence however, is the growing acceptance of the role that Property and Facilities Management plays in the business arena and the influence that physical environment has on business performance".

Linking those that procure facilities with those who operate them:

"It is generally the case that in most organisations, whether they be public or private sector, that the part which procures the asset and the part that is responsible for its maintenance and operation are often organisationally disconnected".

Linking facility management, design and construction within a whole life cycle approach:

"Those businesses which are involved in the provision of the asset must also recognise the impact of the product and premises life cycle on the Facilities Management process and understand the relationship between the Design and Construction process, and the subsequent operations and maintenance phase. To have a sustainable product we must incorporate whole life design criteria into the design brief and clearly understand issues such as life cycle costing and bench marking of quality as well as cost".

Each of these important linkages is explored in this book. The first chapter of the book is introductory. It maps out four rather different and competing directions for the future of facility management. The four parts of the book that follow, explore these four directions in detail. They include edited versions of twenty-four of the Conference papers, with an introduction to each part, highlighting key questions for the future. Those attending the Conference were invited to submit short Speculations concerning facility management futures. Nine of these are included within the fifth and final part of the book. This summarises the dilemmas, opportunities and risks for the future of facility management during the early part of the 21 st century.

Jeffries M., (1999) "Opening Address", Conference Proceedings, *Futures in Property and Facility Management,* (Ed. by Nutt B., McLennan P., Kincaid D.) UCL. London

Acknowledgements

We would like to thank all of those who contributed to the international Conference "Futures in Property and Facility Management", held at University College London in June, 1999. The material from this Conference forms the basis of this book. As with any undertaking of this scale, its success relied on a very large number of contributors. We thank them all. While we would have liked to include all of the Conference presentations, papers, speculations, and a full transcript of the debates and discussions in this book, this was not possible within one volume. We acknowledge and thank all of those who provided their time, expertise and effort and thereby helped us to give access to a wider audience through this book.

First we would like to acknowledge the Conference sponsors, WS Atkins plc, for their support. We are particularly appreciative of advice, time and help from Roger Dyson, and Will Paskins. Secondly we must express our appreciation of the Conference's principal speakers who presented so much valuable material and participated in the lively discussion sessions; Michael Jeffries, Oliver Jones, Roger Reeves, Jeffrey Hammer, Professor Hans de Jonge, Virginia Gibson, Hilaire Graham, Kirsten Arge, Ron Adam, Dominic Treasure, Sir Anthony Walker, Sir Crispin Tickell, Adrian Montague, Michael Medlicott, Jane Herbert, John Thackara, Ashley Dabson, Professor Ranko Bon, Barry Holt and Tim Broyd.

Thirdly we must thank Marilyn Standley, Professor John Worthington, David Kincaid and Phil Roberts who chaired the conference sessions, for their expert moderating skills and their contribution to the Conference discussions.

Fourthly we express our thanks to all of the authors who contributed papers to the Conference proceedings; Stephen Bradley, Geof Woodling, Sue Wood, Derek Worthing, George Cairns, Nick Beech, Wes McGregor, Stephen Brookhouse, Professor Robert Grimshaw, Barry Varcoe, Nigel Oseland, Stephen Willis, Dr John Hinks, Richard Watts, David Garnett, Annemarie Harrison, Christine Landorf, Bertil Oresten, Christine Lofvenberg, John George, James Chilton, Rex and Pat Barnes, Jason Happy, Stephen Bennett, David Kincaid, Edmund Rondeau, Tony Stack, Elizabeth Fox, David Baldry, Professor Tadj Oreszczyn, Phil Roberts, Professor Patrick O'Sullivan, Roderick Rennison, Stephen Brown, Zainy Ali Zeinalabdeen, Bundit Chulasai, Sarich Chotopanich, Dr Stephen Drewer and Harry Bruhns.

Fifthly, a 'thank you' to all members of the FM Exchange at UCL and the alumni of our MSc Facility Management course, for their support. Sixthly we would like to extend our thanks to all conference delegates who enlivened the debate through their questions during the discussion sessions. These transcripts have been used throughout this book to provide the practice perspective which is such an important part of the facility experience. Finally, we would also like to thank the members of the Conference policy and management committees, Jane Bell as the conference adviser, and Wendy Riley and her team as the conference organisers, for their help.

In addition, we owe so much to Michelle Julier for her administrative support throughout the development of the book, to Joanna Saxon for creating the manuscript templates and to Caroline Woodbridge for preparing the drawings and tables and to Juliette Baigler for proof-reading and corrections. Many thanks for your help. Finally, we would like to thank Madeline Metcalfe and staff at Blackwell Science for their expert guidance during the final preparation of this book for publication.

Professor Bev Nutt
Peter McLennan

1 Four Trails to the Future

Bev Nutt, University College London.

The Strategic Objectives of FM

At a national level, the strategic objective of facility management is to provide appropriate infrastructure and logistic support to business and public endeavours of all kinds and across all sectors. At a local level, its objective is the effective management of facility resources and services to provide shells of support to organisations, their operations, their working groups, project teams and individuals, and to their suppliers and customers. So overall, the primary function of FM is resource management, at both strategic and operational levels of support. From this perspective facility management may be defined as "the management of facility resources and services in support of the operations of an organisation" (UCL, 1993).

The supporting role of facility management is perhaps the most singularly significant factor that distinguishes it from business and operations management generally. This support role is wide-ranging. It impacts on the financial issues of facility investment, asset value, and operational costs and benefits; the human issues of purpose, use, environment, security, safety and health; and the physical issues of space, structure, technology and maintenance. But most significantly, it covers the management structures that link knowledge and experience across these financial, human and physical areas of concern.

Figure 1   Four Generic Trails to the Future.

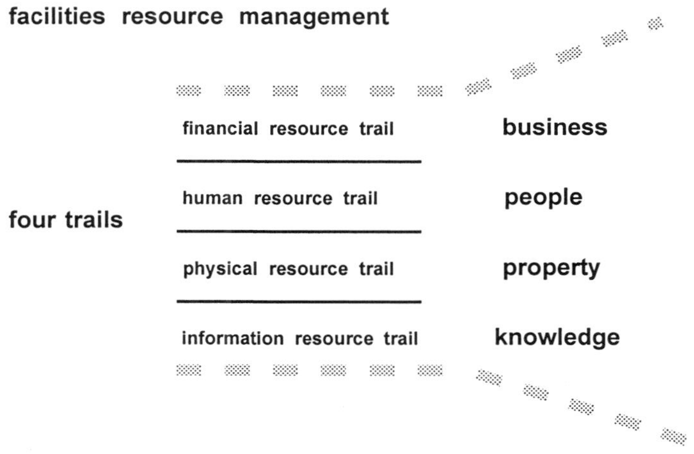

facilities resource management

four trails	financial resource trail	business
	human resource trail	people
	physical resource trail	property
	information resource trail	knowledge

Within the resource management context, this introductory chapter sets out to explore four basic 'trails' to the future, concentrating on strategic rather than operational directions. The four trails correspond to the generic types of resource that are always central to the FM function as set out above; the management of financial resources (Business), human resources (People), physical resources (Property), and the management of the informational resources (Knowledge). Each of these trails, as shown in Figure 1, will be considered in turn, with speculations on the directions that they might promise for the future. The following four parts of this book examine the dilemmas, opportunities and risks for the future that each of the four trails might entail.

Before setting out along these trails, the strategic purpose of the trip needs to be made clear. For more than thirty years the idea of strategy has been developed within the military, political, business, planning, design and management fields (Nutt, 1988). The essence of a strategic approach is making decisions in changing, uncertain, unpredictable and competitive circumstance. In 'defence mode' it is about preserving existing options, maintaining flexibility, generating new options, preparing contingencies and the means of response, with intelligence systems to monitor changing conditions. In 'attack mode' a strategic approach is reliant on flexible response, active awareness of the capabilities and limitation of all sides, rapid option appraisal and decisive choice of action, all directed to increasing the room for manoeuvre against the 'enemy' and achieving operational success. Facility management faces 'peace time' situations of this kind.

Figure 2   Risk and Opportunity

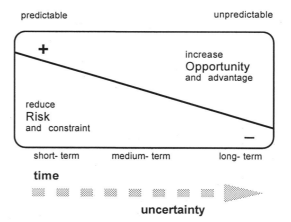

Management strategists know of the dangers of attempting to predict future events precisely. Forecasting the future more than a few years ahead is normally impossible. All that can be done is to guess at the likely directions of change. Strategists therefore expect that the further they look into the future the more uncertain they will become, moving from a predictable 'short-term' to an unpredictable 'long-term' position. The level of uncertainty increases with time. Figure 2 shows this basic position. Short term futures (1-2 years) can often be predicted. Over the medium term (2-5 years) decision makers must rely on trends and forecasts. In the longer term (5-10 years and more) the level of uncertainty increases exponentially, with the future becoming highly unpredictable at best, unknowable at worst.

The strategic approach is therefore directed to the management of uncertainty over time. But how do you manage uncertainty? Typically, within the context of facility management, it has two complimentary objectives. One objective is negative the other positive. On one hand, property, facilities and support services can inhibit the goals and work of organisations, their teams and individuals. On the other hand, they can contribute to productivity, human effectiveness and well-being, acting as a catalyst for change and facilitating business success. So in looking to the future negative objectives (close down risk : avoid failure) and positive objectives (open up opportunities: achieve success) need to be combined. Figure 2 illustrates these two positions. The first objective focuses on measures to contain, reduce, transfer and avoid risks and constraints, both known and unpredictable, that property and facilities can impose on an organisation and its employees. The second seeks to maintain and generate opportunities and advantage, both planned and fortuitous, that property and facilities might provide. In facility management the negative role is of paramount importance. It is only when this is securely in place, that the positive contributions of FM may begin to be realised.

Risks can be assessed in terms of:-

• The types of potential risk of all kinds, both 'hard' and 'soft'.
• The significance and degree of risk within each area, both quantitative and subjective.
• The effectiveness of existing risk management arrangements.
• The contingency measures that are in place to respond to unpredictable risks should they occur..

Similarly, opportunities can be assessed in relation to:-

• The areas of opportunity that are available.
• The number and significance of options within each area.
• The measures that are in place to search for new opportunities.
• The intelligence arrangements that are in place to identify unpredictable opportunities should they arise.

So along each of the four trails shown in Figure 1, a strategic evaluation of future risks and opportunities may be undertaken in as rigorous a way as circumstances require, looking to reduce risk and increase opportunity into the future.

Direction 1 :
The Financial
Resource Trail

(BUSINESS)

The finance trail will be the first to be explored. Traditionally it has had three pathways; the first directed to property investment decisions, the second to the management of property assets, the third to the management of facility operating costs, all within the context of the property market which tends to be the most illiquid vehicle for investment. Over the last fifteen years the financial resource trail has been dominated by a simple business imperative across most sectors, resulting in downsizing, outsourcing, reduced operating budgets, property consolidation, disinvestment and disposal. Facility management has been part of a 'cost-cutting' culture for short-term business benefit and shareholder value. The finance trail has been a capitalist trail. But once inefficiencies, under utilisation and waste have been squeezed and managed out, what then? Reductionist measures of this kind certainly produce 'balance sheet' improvements but they cannot be continued indefinitely. Further downsizing and cost-cutting will begin to harm the very operations that they set out to support.

In the UK, the introduction of the Private Finance Initiative (PFI) and Public Private Partnerships (PPP) is changing these priorities fundamentally and fast (HM Treasury, 1995). The finance trail has become truly pioneering. PFI contracts look to identify, allocate and manage financial risk over the working life-cycle of a facility. It is the operational value of infrastructure and facilities that are at the centre of concern, targeting the output needs of organisations, their staff and customers over a twenty-five to thirty years life-cycle (CIC, 1999). In PFI projects the brief is defined through an 'output specification' that clarifies support requirements, operational performance and property strategy. This PFI provision alone is changing the future of property and facility management and could reverse the priorities of the past by:-

- Promoting the 25-30 year operational FM phase to a position of primary importance, demoting the 2-3 year design and construction project phase to merely a 'means to an end'. This dramatic reversal of emphasis is likely to be especially traumatic for the design team.
- Replacing the traditional design project brief with the operational FM brief as the main driver for development, bringing a new reality to the traditional roles of the property professions.
- Developing a comprehensive 'effective-life' framework for facility finance to enrich simple discounting and life-cycle costing approaches.
- Prioritising 'use value' rather than 'asset value' generally.

These innovatory but politically sensitive steps along the finance resource trail threaten a break with the past and could have a profound impact on the profile and importance of facility management. In the USA the DBFO (Design - Build - Finance - Operate) approach promises to make a similar impact. But many questions remain unanswered. How should the target 'life-expectancy' of a facility be determined at a given location within a particular sector? To what extent should increases in life cycle costs be anticipated in facility budgets as a direct result of accelerating rates of organisational change?

The allocation of financial risk is a central concern in PFI projects. To date, a legalistic and risk adverse 'accountancy' type of approach tends to have been adopted, concentrating on the 'hard' and 'knowable' short-term risks. Overall, the financial trail seems to be moving from a 'cost-cutting' past to a 'risk adverse' future. This is itself risky. In the longer term, it is the less predictable and 'softer' risks that are often the more threatening. For example, it is inevitable over the life of a PFI contract that new operational procedures will be put in place, functions and uses will change, technological innovation will continue, different business demands will arise, and new market opportunities will evolve. All can induce early functional and locational obsolescence, reducing or eliminating the 'use value' of the facility. The origins of functional obsolescence pose a far greater risk to financial viability than the physical causes of facility obsolescence and cannot be ignored (Nutt and Sears, 1972).

But the consideration of risk is only half of the story. Life-cycle opportunities, as indicated in Figure 2, have hardly begun to be considered from a financial perspective within the PFI scheme and surely warrant more attention in the future. A more comprehensive and rigorous approach to the identification and management of both risks and opportunities needs to be put in place. Developments here would help to consolidate the new directions for FM along the FR trail, with convergence between the concerns and priorities of operational management and property management to create a balanced and expert 'real-life' approach to facility finance for the future.

Direction 2 : The Human Resource Trail

(PEOPLE)

The second trail to be explored - the human resource trail - could become the most revolutionary route to the future. The consensus of the late 90's is that fundamental changes to the nature and organisation of work are underway, pushed by wave after wave after wave of IT innovation. Today we are witnessing ever shorter business time horizons, the diversification of work practices, more responsive working arrangements, the global dispersal of work, new multi-venue and multi-location ways of working, all with increased reliance on subcontracting and partnering. These developments are part of the fashionable notion of 'flexible working'; work that is 'time flexible', 'place flexible' and 'location variable'.

These changes are altering facility requirements fundamentally. The main FM objective on the HR trail is to support the effective deployment of human resources. While good facility management can enhance the performance, productivity and well-being of individuals and teams, it has no remit to manage human resources directly. So the contribution of FM to HR management will tend to be subtle and indirect.

Flexible working will demand an agile approach, both to the acquisition, deployment and release of human resources, and to the provision and management of the support facilities that they require (Pettinger,1998). The beginning of the end of the dedicated workplace seems to be at hand, whether it be the executive office or the individual computer workstation, to be superseded by a diverse range of working venues.

Flexible working poses major risks for both organisations and individuals alike. Organisations will have to build their corporate cultures of enterprise, motivation and loyalty across the barriers of multi-location, multi-venue and multi-time working arrangements. For the individual the problem will be to maintain commitment, confidence and work performance in diverse and often isolated work environments. The development of completely different attitudes to organisational culture and leadership will be required to support the new ways of working. These will need to be 'output' driven but humanistic, with emphasis on self-motivation and self management.

But what do these directions imply for the future management of property, facilities and the markets in which they operate ? At its most mechanistic, the human resource capacity of an organisation simply consists of the total people-hours per annum that are available to its business. The total human resource can be broken down by type of skill, knowledge, and experience, producing HR time budgets at organisational, section, team and individual levels, for deployment in place, space and time. The strategic allocation of this range of 'people-hours' of resource generates the profile of demand for FM support, again distributed in place, space and time. The increasingly varied and volatile nature of these demands is already making some 'best practice' redundant, particularly in relation to space planning and management. The explicit management of 'availability', 'time' and 'utilisation' are likely to be the major problems for facility management from the HR perspective.

The Human Resource Trail will involve the management of 'place-flexible' and 'time flexible' facilities, probably moving towards a form of Logistics Management (Tripp, 1991), to provide flexible 'bottom-up' and 'real-time' support to individuals and their tasks, sectional operations and their project teams, divisions and their missions, organisations and their business objectives. Extensive expertise in the management of flexible response support systems and logistics already exists in the military field. Logistics systems of this kind rely on secure procedures to support the management of operations at all levels. For operational viability and success, a secure quantitative framework of support is essential within the logistics management approach. Reliance on 'soft' and 'touchy' management approaches alone, will not be sufficient. Here facility management will need to become accountable for:-

- Operational Capability : in delivering facility capacity (size), availability (time) and flexibility (change) of the types and locations (variety) of facility supports that are required.
- Contingency Provision : planning 'what-if' arrangements for response to unexpected events, should they occur.
- Operational Performance : monitoring the impacts of facility and service support systems, both positive and negative, on people, their operations and performance.
- Operational Effectiveness : evaluating the affect of facilities and support services on work output, personnel satisfaction and organisational success

Facility management on the human resource trail will need therefore, to move closer to core business strategy with people, facilities and support services being seen as a key business resource to be dynamically deployed for long term advantage, rather than as a cost to be cut for short term gain. With this perspective FM strategy could become an integral and essential part of HR strategy itself. It seems likely that the scope of FM will need to widen considerably, with delegation of some FM functions to individuals and working teams. The 'help desk' of the past will evolve to a 'help-interface' of the future. FM responsibilities could widen to include transport and communication support arrangements, information management, and perhaps extend to help with wider community, social, leisure and family support arrangements.

Figure 3   The People Trail.

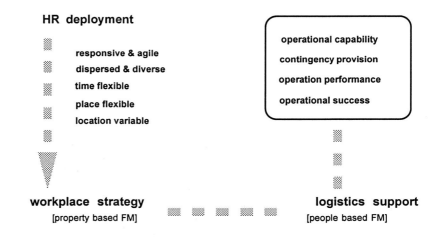

So the realities of the HR trail argue for a basic shift from a 'property-based' to a 'people-based' focus for FM, as summarised in Figure 3, with 'mobile' rather than 'place-fixed' workplace strategies. The Trail will need a combination of 'hard' and 'soft' management approaches to provide logistic support at individual, team and organisational levels, requiring 'task-force' support management, both on war-time 'alert' and peace-time 'steady-state' footings.

Direction 3 :
The Physical
Resource Trail

(PROPERTY)

The third trail - the property trail - is the most predictable trail to the future. The physical inertia of the built environment makes it so. Each year some 1.5% of the UK's building stock is demolished, mainly to be replaced by new buildings. A further 2.5% is subject to major refurbishment and renovation. In the rest of Europe the figures are much the same. Therefore, in any year only about 4% of national facility stock will be changed. So on the first part of the property trail, business and public initiatives of all kinds will be reliant on the support of buildings and infrastructure that have been inherited from the past. It will only be towards the second part of the trail beyond 2010, that new types of facility support and innovative design could begin to make a significant impact on business performance overall.

The early stage of the property trail has two linked paths through which supply side improvements can be achieved:-

- through the better management of the existing facility stock
- and its physical modification and adaptation.

Here management measures will include:-

- Utilisation Strategies for managing both space and time more effectively: even greater emphasis on the intensive use of property and facilities, requiring expert procedures for utilisation management.
- Rationalisation and Disposal Strategies: further efficiency gains from workplace rationalisation and innovation, the introduction of tougher building performance regimes, more effective FM arrangements and a more sophisticated approach to the disposal of property and parts of buildings that are no longer required.
- Flexible Tenure Strategies: responding to the needs of business for highly flexible and shorter tenure arrangements, providing choice and opportunity through mixed tenure combinations, including long-term ownership, coupled with medium-term leasing and the short-term support of serviced space and temporary facilities to meet the highly volatile episodes of demand.

Figure 4   The property trail.

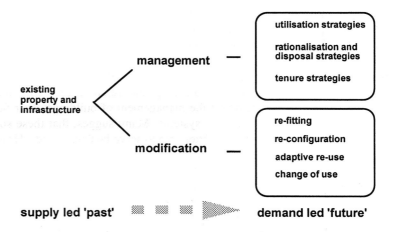

Over the last ten years there has been an excess in the supply of built space in many sectors and at many locations. Further reductions in the aggregate demand for space per employee or per unit of production should be anticipated, with priorities directed to:-

- Physical Modification Strategies: today's patterns of demand are testing the ability of buildings to accommodate radical change as never before. The frequency of modification and re-fitting will need to increase significantly.
- Adaptation and Reconfiguration Strategies: the selective demolition of parts of the existing stock to permit fundamental reconfiguration and renewal, providing new opportunities for re-use and mixed use will increase (Nutt, 1997).
- Change of Use Strategies: the excess supply of built space will increase the rate at which redundant space needs to be converted to support new classes of use and mixed use.

All of these tendencies, as summarised in Figure 4, are likely to intensify on the second part of the property trail. But a re-appraisal of the purpose, form and life cycle characteristics of buildings will also need to be undertaken in order to move away from the introverted culture of the construction industry.

For example, the recent Egan Report set out seven improvement target for the construction industry in the UK (DETR, 1990). These targets aim to reduce construction time, capital costs, construction defects and site accidents, and to improve 'on-time' completion rates, productivity and construction industry profitability. While these are laudable aims, they are all for the short-term benefit of the construction industry and its immediate clients. None of the targets are directed to product innovation for use and management, or to the operational requirements of the property consumer, or to the national need for better property and infrastructure support for its public and business endeavours. A wider review beyond the narrow interests of the construction industry is long overdue to anticipate the likely impact, both of changes in demand and changes in technology, on the future of facilities towards 2020.

Over this period the rate of technological innovation seems set to accelerate into the future. The likely impact of this technological change on the production of buildings over the next twenty years or so, is open to debate. There are two contrasting viewpoints. The first suggests that technological advances will lead to even greater reliance on highly integrated and sophisticated systems for the management of buildings, their fabric, services, equipment and operating systems. Many suggest that these so called 'intelligent buildings' will be better able to serve business needs. However, as the shelf-life of technology shortens, so the ease with which it can be upgraded or replaced becomes of critical importance. Complex embedded technology can increase the risk of early obsolescence.

The second and contrasting view is that the emerging compact, mobile and personal computing and communication systems, voice activated and less reliant on 'hard wiring', will permit even greater individual choice of work-style, working venue and location. It is suggested that the 'footloose' nature of this next generation of IT work support systems, will enable many work, retail and educational activities to become somewhat more independent of buildings and services than in the past. This in turn may continue to shift the balance of concern during facility design towards the development of high quality 'human', 'green' and 'social' building environments but with a decreasing need for sophisticated and intelligent systems to be integrated within the physical building fabric. Buildings would become simpler. The interface between technology and buildings could loosen, resulting in less 'intelligent' but more robust and flexible facilities overall.

These two contrasting veiwpoints, 'intelligent high tech' as against 'simpler and greener' buildings, are not necessarily incompatible one with the other. A mixed approach can often be adopted. Here there are two key questions. First, how robust is the level of technology for facing the possible patterns of change in function and market condition? Second, what is the probable shelf-life of the embedded technology and how easily may it be adjusted or upgraded in the future? These questions relate to the risks of 'over' or 'under ' specifying the level of technology to be incorporated within any specific development. The selection of an appropriate 'high tech', 'low tech' or hybrid strategy is essential.

Overall, the range of potential areas for product innovations for improved use and manageability are vast, but innovations might be expected to include:-

- Versatility: today's 'use classes' and traditional building types will provide an inadequate basis for re-use, mixed use and changes of use in the future. The validity of designing for a particular building type within a single class of use in a fast changing world, needs to be re-examined.
- Re-differentiation: the ability to fundamentally redistribute built space and to reconfigure building structure and sub-systems, could improve the adaptability potential of buildings and their capacity to accommodate radical change.
- Diversification: greater variety of building stock types will be needed to support the diverse and widening range of business and social demands.
- Robustness: buildings with robust structure, spaces, fabric and services, reliant on high quality but low technology solutions, will be intrinsically more adaptable than those with highly sophisticated technology.
- Innovation: new generic types of building stock, financial systems and tenure arrangements, will need to be created to support changing multiple uses and different mixed use combinations, over time.

Directions such as these will be important to change attitudes within the construction industry and begin to move from an 'introverted' past towards a more responsible and 'responsive' future. The management and design implications of issues such as these are considered next, along the last of the four resource trails.

Direction 4 :
The Knowledge
Resource Trail

(INFORMATION)

At the start of this trail everybody faces three very uncomfortable questions. What is the shelf-life of their knowledge and experience? Can it be extended? How soon will it become worthless? In today's world the answers for most individuals are likely to be 'short', 'seldom' and 'soon'.

Facilities Management expertise is particularly vulnerable. Its knowledge base is at a primitive stage of development, its terrain largely unexplored.

While the relevancy and promise of FM has become recognised by business, industry and government, it remains reliant on management and technical knowledge that has been plundered from other fields. FM is now challenged to make a distinctive contribution to management expertise in order to build a secure knowledge platform to support future developments. If its shelf-life is to be lengthened, then it needs to build an expert knowledge structure of its own to move beyond best practice and to bridge the gap between promise and performance.

The FM knowledge trail covers an extremely wide territory, only a small part of which can be considered here. It has three main origins on which to build; knowledge of property, general management knowledge, and knowledge of facility design and facility management. The first two of these are relatively secure, the third is undeveloped. It is the interface between facility management knowledge and facility design knowledge, at the strategic level, that needs to be addressed urgently, to understand the impact of:-

- Management on Design: what management considerations should inform the design process? What management opportunities are required, what constraints to management should be avoided?
- Design on Management: how will any given facility design affect its management when in use ? What opportunities will it offer, what limitations might it impose?

Knowledge here will help to ensure that facility objectives support organisational objectives, that facility provisions are aligned with end-user needs, and that design concepts are compatible with property strategy overall. A knowledge base needs to be developed at the interface areas where the concerns of management and design always overlap, during life cycle briefing, life cycle design and life-cycle management. Each will be considered briefly in turn.

Life Cycle Briefing

In the past, the briefing process has rested on an assumption that the needs of the client, the function of a facility, and the requirements of the 'first-hand' user are an appropriate starting point for design. This is not a secure position. Firstly, the needs of the client organisation, end users and market conditions, can no longer be forecast with confidence beyond a short two to five year time horizon. Second, the step by step logic of the traditional briefing process, as incorporated in the RIBA Plan of Work, is inadequate to meet the diversity of client needs in today's climate of continuous innovation and change. Thirdly, conventional briefing methods encourage customised rather than generic approaches. As a result new building stock tends to meet the specific needs of a particular client at a particular time, rather than the operational needs of businesses over the medium to longer term. So the demand-led briefing process, targeted on the client's corporate objectives, the requirements of the first generation of users, contemporary working practices and current market conditions, is becoming obsolete. A less tailored and more generic approach to facility briefing is required.

The case for a full life-cycle framework for briefing that covers both management and design decisions, has been made before (Nutt, 1993). The recent introduction of the PFI has added weight to the arguments made. The three parts of the life cycle briefing process are shown in Figure 5. The first and major part of the cycle relates to the continuous briefing process for the management of facility resources and services to support the changing operational needs of an organisation over time. This informs the second part of the cycle concerning strategic briefing for business infrastructure support within the corporate property strategy overall. Finally, and only very occasionally, the briefing cycle will include a third part, a project briefing loop, as and when new buildings are required, or when change of use is planned, or when the existing facilities are to be refurbished.

Figure 5   Life cycle briefing.

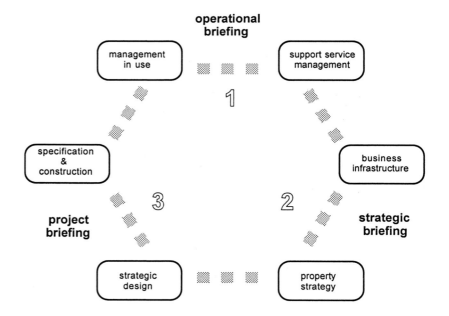

The development of this life-cycle briefing process is an exciting and challenging area for facility management. It will cover the continuous adjustment, modification, refitting and redirection of operational support systems over time to meet the changing needs of the corporate client and end-user. At the strategic level it will include the reconditioning, expansion, contraction, realignment and adaptation of an organisation's facilities and the repositioning of its support services to meet emerging business needs and new priorities as the future unfolds. The life-cycle brief will need to be underpinned therefore, by a secure knowledge base concerning the generic operational needs of organisations and, for practical implementation, it will be reliant on the development of facility management procedures for anticipating, monitoring and responding to changing operational requirements. It will need to feed forward this knowledge, first to inform property and infrastructure decisions within the strategic briefing process, second, to inform the early stages of any design and construction brief. Finally, this life cycle briefing approach will begin to provide the data structure that is so essential to the linkage of management to design.

Figure 6   Life cycle design.

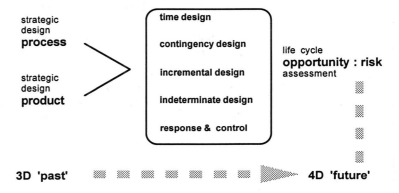

Life Cycle Design

A strategic approach to life cycle design has two objectives. The first is to make the process of design more strategic, providing a more intelligent, reliable and accountable design service. The second objective is to make the product of design more strategic, improving the viability of facilities in relation to their dynamic use over time (Nutt, 1988). Both are reliant on the life cycle brief, as sketched out above, to ensure that operational issues become a paramount concern of design, informed by the FM team acting as the 'intelligent client'. Design and management policies will probably include:-

- Time Design: The consideration of time underpins the strategic approach to life cycle design. The time frames for facility use and the life expectancies of its constituent parts need to be explicitly addressed. Wherever possible 'short life - high change - low cost' elements and 'long life - low change - high cost' elements should be separated both in design and management systems.
- Contingency Design: In addition to meeting known short-term requirements, contingency measures need to be 'designed-in' to face longer term uncertainties and to reduce the risks of functional obsolescence. Contingencies commonly include the over-provision of servicing capacity, some structural redundancy, the oversizing of spatial provisions and strategic measures to permit change of use and change of function.

- Incremental Commitment: The design decisions of today should not limit the future. In areas where there is high uncertainty about long term requirements, positions need to be adopted that preserve the freedoms of future users and managers to take action at the appropriate time.
- Indeterminate design: A very old but 'modern' idea from the 60's, arguing that the tendency to design to meet corporate objectives, clear goals and the precise brief, within a single property strategy, should be avoided (Weeks, 1963). Design ingenuity is needed to maintain a degree of ambiguity to facilitate future change, re-organisation and management choice.

Client organisations should expect measures of this kind, as summarised in Figure 6, to be examined routinely so that the robustness of any design proposal is assured. Design concepts for facility proposals have not, in the past, been thoroughly examined and comprehensively assessed in relation to the risks and opportunities that they may hold for the client during their subsequent use and operation, particularly in regard to their manageability, adaptability and disposability, once constructed. The introduction of the PFI has highlighted the lack of formal risk assessments here. Practical techniques for full life-cycle 'Opportunity : Risk Appraisal' (ORA) need to be developed so that the decisions of design can be scrutinised for reduced risk, increased opportunity and for improved robustness and sustainability overall.

Life Cycle Management

In the past, buildings have tended to be seen as passive products, not as active support environments with built-in measures for management, control and redirection. Measures to improve manageability and flexibility are perhaps the most important issues here. For many years designers and managers have recognised the importance of flexibility. But often the so-called 'flexible design' has only been so on paper. In reality, few designs offer strategic flexibility to the client in support of their business, or operational flexibility to the facilities management team. The concept of flexibility needs to be widened considerably, particularly from a management perspective. In the past, the issue of manageability has not been a major concern of designers. Today we need creative design solutions that generate new options and choice for management. New forms of design knowledge are needed to help improve the fundamental responsiveness of facilities to the pressures of unremitting change.

Design for improved flexibility and manageability will help to converge management and design priorities, promising to transform the ways in which property and facility management is currently conducted. Life-cycle management needs the support of:-

- Use Flexibility: increasing the capacity of space to support changing patterns of use, design for combinational use, and design to permit progressive shutdown, seasonally, daily, hourly.
- Operational Flexibility: increasing the capability of a facility to respond to changing operational requirements without physical change. This is perhaps the most valuable type of flexibility for FM.

- Physical Flexibility: design for physical modification, re-fitting and the replacement of plant, services, components and IT systems.
- Property Flexibility: design for adaptation to increase the ways in which property can be subdivided into a number of separate buildings, design for selective demolition and core reconfiguration, through which radical adaptation can be achieved.
- Market Flexibility: design for improved marketability, increasing the capacity of the design to accommodate different types of organisation and uses, improving subletting possibilities and the ease of property disposal.

Figure 7   Life cycle management.

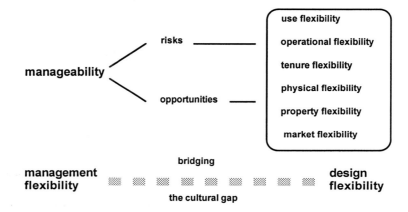

These issues are summarised in Figure 7. In the past most facilities have been designed for a given client and given class of use, but flexibly with the probabilities of future change in mind. The logic of this process may need to be reversed. In the main, facilities will need to be designed for multiple categories of use, but flexibly so that they can be customised to meet the needs of the 'first hand' user.

Nine Strategic Positions

There are nine initial positions from which each of the four trails into the future might begin. These positions are shown in Figure 8. They range from the 'exposed' position (increased risk - reduced opportunity) which is logically foolish unless forced, to the 'robust' position (reduce risk - increased opportunity) which is logically ideal. It might be expected that the aim on all trails will be to move to the 'robust' position. This is not always so. In some circumstances it is wiser to adopt 'insecure' or even 'exposed' positions in the short term, gambling on longer term advantage. For example, most innovations are risky in the short term, but can be hugely rewarding thereafter. However, in general over the longer term, the logical position is to reduce risks wherever possible, while opening up new opportunities of all kinds. So from any initial position in Figure 8, the ways in which it might be possible to move from the 'bottom left' towards the 'top right' should be explored.

Four final examples will be considered. First, in the case of the financial resource trail, its PFI branch starts from the 'neutral' position shown in Figure 8. Facility briefing and financing in PFI projects, has tended to be developed from a conservative position that 'contains' known risks and maintains 'traditional' opportunities. The approach has yet to set out, in the main, to prepare for unforeseeable risks and to generate new opportunities for the future.

Turning to the human resource trail as profiled in this chapter, there can be little doubt that it sets out from the 'brave' position. New ways of working promise vast opportunities, particularly in the office, retail, education and residential sectors. They also involve new types of risk for businesses and individuals alike. So strategies on the early part of the HR trail must attempt to move from the 'brave' position of increased opportunity and increased risk, towards the 'secure' position where opportunities are preserved while the new forms of risk are contained and reduced.

Figure 8   Nine Strategic Positions.

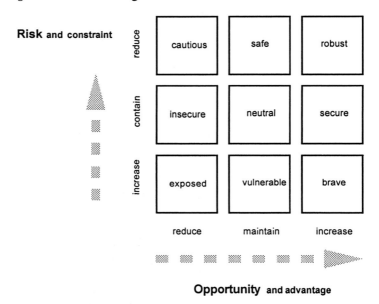

The third example concerns facility management practice which has also tended to adopt a conservative approach. Generally, it has worked out of the 'neutral' box, relying on bench marking and best practice for progressing its position. In this its priorities have been largely risk adverse. It has yet to create its own distinctive management theory and practice and is challenged to move towards the 'secure' position by opening up new opportunities for the future support of organisations and their staff.

The last example concerns facility design, which over the last twenty years has tended to operate out of the lower left hand side of Figure 8. At best the creative attitudes of design have had neutral effect on operational risks and opportunities. But creative design often generates its own breed of risks, pushing corporate clients into the 'exposed', 'insecure' or 'vulnerable' boxes. At worst, it has imposed an 'art albatross' on organisations with buildings that increase risk and diminish opportunity.

So the current approaches to both facility management and facility design may be characterised by positions on the bottom left hand side of Figure 8. Hopefully, with appropriate knowledge support, future facility design and management strategies will target the top right hand side, aiming for 'safe', 'secure' and 'robust' positions for FM. The future paradigms for management and design must be dynamic. They must focus on the fourth dimension, the dimension of 'time', 'change' and 'uncertainty, so that facilities are planned strategically, designed strategically and may be managed in a more strategic fashion.

This introductory chapter has briefly explored four resource trails, which basically have been about Business, People, Property and Knowledge. All seem likely to be subject to radical change towards 2020, with a wide range of possible futures. Parts of the four trails might merge or diverge, splinter or spawn new pathways. While the trails share a common objective - to provide strategic and operational support to all of our endeavours - they work to different agendas, serve different interest groups, with conflicting priorities and ambitions. The four trails are in competition one with another. Rather than attempt to force a single integrated future, facility management should build positively on the creative tensions between the four resource trails. It should work to create a platform for innovation that accepts diversity, variety and ambiguity, welcoming an expanding range of alternative facility management futures. This diversity and variety will be explored in detail along each of the four trails to the future.

References

Construction Industry Council (CIC), (1999) *Construction Key Guide to PFI,* Thomas Telford, London

Department of the Environment, Transport and the Regions (DETR) (1998) *Rethinking Construction*, (The Egan Report), HMSO, UK.

H.M.Treasury, (1995) *Private Opportunity, Public Benefit: progressing the Private Finance Initiative,* London

Nutt B. and Sears S. (1972) Functional Obsolescence in the Planned Environment, *Environment and Planning*, Vol. 4, 1972, pp. 13-29.

Nutt B. (1988) The Strategic Design of Buildings, *Long Range Planning*, Vol. 21, No. 4, pp. 130-140.

Nutt B. (1993) The Strategic Brief, *Facilities*, Vol. 11, No. 9, pp. 28-32.

Nutt B. (1977) Adapting and re-using buildings, *International Journal of Facilities Management*, Vol. 1, No. 3, pp. 113-121.

Pettinger R. (1998) *Managing the Flexible Workforce*, Cassell, London and Washington.

Tripp R. et al. (1991) A Decision Support System for Assessing and

Controlling the Effectiveness of Multi-Echelon Logistics Actions, *Interfaces 21*, July-August, pp. 11-25.

University College London (UCL), (1993) *MSc: Facility and Environment Management*, course definition.

Weeks J. (1963-64) Indeterminate Architecture, *Transactions of the Bartlett Society*, Vol. 2, pp 85-106.

Part One

The Business Trail

2 Introduction to the Business Trail

One of the most common claims made by facility directors and managers is that FM is of crucial importance to business and that it makes a significant contribution to business success. But what is the reality? What is the viewpoint of business organisations themselves ? Unfortunately there is scant evidence that facilities management is part of the strategic agenda of most business organisations. It is certainly invisible in the curricula of the leading UK and North American Business Schools. While property and business infrastructure management issues are included in a few curricula, they tend to ignore the specific contributions that those in the field of facilities management claim that they make. So what is the potential linkage between facilities management and core business activity? In principle facility management might contribute to, or detract from business performance in a number of ways, with its performance assessed in relation to:-

- its contribution, or not, to the objectives of an organisation and to business success;
- its support, or not, to business strategy and operations;
- its contribution to improving productivity of the business;
- its management of the assets of an organisation's property portfolio;
- its impact on facility operating costs;
- the effectiveness, or not, of its own facilities management policies and procedures;
- the quality of the working environment;
- the delivery and quality of out-sourced, part-sourced and in-sourced services;
- the support it provides to the employee and end-user;
- the quality of service as received by the customer or consumer.

These areas of FM performance operate at different levels of business support, on a variety of time-frames, to the benefit of different stakeholders, often with conflicting objectives and priorities. For example downsizing and cost cutting to improve competitiveness, corporate business performance and shareholder benefit, can have negative impact on employee morale, well-being, corporate loyalty and customer service. But once inefficiencies, under-utilisation and waste have been squeezed and 'managed out', what then? Here further downsizing and cost-cutting is likely to have a negative rather than positive affect on business performance, hurting the interests of those that FM sets out to support. So the potential impact of FM performance is a highly complex issue. It is not surprising therefore, that there is at present, a gap between the promise and performance of facility management.

Recent innovations in the ways in which property and facilities may be procured and financed, has added to this complexity. In the UK, the Private Finance Initiative (PFI), re-branded as Public Private Partnerships (PPP) has introduced a new vehicle and arrangements for procuring infrastructure, facilities and their associated support services, with the involvement of both Private and Public Sector finance (H.M. Treasury, 1995). While the UK has pioneered public-private partnership developments, similar initiatives are now being undertaken around the world (Economist Intelligence Unit, 1999).

Adrian Montague, the Chief Executive of the UK Government's Treasury Taskforce, gave the keynote PFI Address at the 'Futures in FM' Conference (Montague, 1999). He outlined the progress made and described the challenges still to be faced by PFI as it matures and gave his views of the way ahead. By 1999, over 250 PFI contracts had been signed in the UK to a total value of more than sixteen billion pounds (HM Treasury, 2000). The continuing success of the programme is evidenced by the current level of thirty to forty PFI deals concluded each year, triggering some five billion pounds of capital investment since the present UK government took office in May 1997. The range of applications has been wide. It has included infrastructure projects; roads, bridges, railways, airports, utilities, and facility projects; prisons, hospitals, universities, schools and now, social housing and urban regeneration schemes.

All PFI projects require a balanced sharing of risk between the parties, with a transparent demonstration of better value for taxpayers money through the PFI scheme when compared to the conventional capital spending that would have been required to create the same facility with public sector funding alone. PFI contracts look to identify and manage risk over the working life of a facility with the operational value of the facility as one of the centres of concern. The approach links private sector risk management expertise with the potential for facility design and facility management efficiencies, by combining finance, design, construction and operation within a single package.

From a facility management perspective, these developments open up some exciting prospects. Four specific opportunities should be mentioned by way of introduction here. They relate to facility use, facility cost, facility risk and facility value.

First, PFI projects are based on partnership arrangements between the client organisation, those who are to finance, design and construct the facility, and also those who are to manage, maintain and operate the facility in use and deliver support services over a typically 25-35 year concession period. PFI projects therefore require the input of facilities management knowledge and experience within the process of facility briefing, output specification, design, construction and commissioning. This places facilities management in an important new arena of involvement and influence.

Second, in relation to cost, PFI ensures that not only capital costs, which average only some 35% of total costs (HM Treasury, 2000), but also the huge operational costs of property, facilities and their support services are explicitly recognised and addressed within the investment strategy, again emphasising the importance of facility management involvement during the project stage.

Third, the central concern in PFI projects is an allocation and transfer of risk to the parties of the partnership. All identified risks must be included, both those arising within the design and construction period, and the operational risks over the concession period. But this is not happening. While the National Audit Office recommends that all known risks are included (NAO, 1999), there is clear evidence that only financial and construction project risks seem to have been routinely part of PFI bids to date (HM Treasury, 2000). The consideration of operational risks has hardly begun. The next stage of PFI implementation should therefore involve a comprehensive and systematic examination of operational risks over the facility's life-cycle.

Forth, in relation to value, and largely as consequence of PFI innovations, the concept of business accommodation and service delivery is changing rather radically. Property is beginning to be considered more as a business support service than as a financial asset, helping to free core business capital while reducing the costs and increasing the quality of service delivery. As a result a basic shift of emphasis seems to be underway in facility investment criteria and in asset management generally. Under PFI the 'operational value' of a facility over the 25-35 year period and beyond, has to be considered alongside the traditional business 'asset value' of the property for disposal or redevelopment. It seems that a new balance is being struck between the 'use value' and 'investment value' of facility assets, with priorities shifting away from 'asset value' towards 'use value' overall.

It is too early to assess the long term impact of changes such as these. But they do promise a number of new opportunities for facility management involvement along the business trail. Key questions for the future include:-

- How can the linkages between FM and business be demonstrated and quantified?
- Will the distinction between 'core' and 'non-core' business activity be useful or valid in tomorrow's business world?
- What market niche might FM occupy within business management in the future?
- How can FM develop a strategic capability that directly contributes to business performance and success?
- How can the contribution of facility management to improving business productivity be measured and assessed?
- How can the most appropriate support services strategy be determined for a given business organisation?
- Will FM skills of the future need to include essentially core business skills?
- How can the value of FM be marketed internally and externally?

This first part of the book, along the business trail, concentrates on questions such as these and the issues that they raise. Chapter 3 looks at contemporary changes to the ways in which facilities are being provided, the integrated management of property and services, and the diversification of the business support environments that are being created. The diversification of the rules governing the property market, coupled with an increasingly sophisticated range of outsourcing arrangements, promises to provide a much greater variety and more flexible set of business support environments for the future. Chapter 4 examines these innovations from a supply-side perspective. It provides a detailed account of the Principal Finance approach, through a case explanation by one of the most successful providers of comprehensive financial, property, facilities and services support. The aim is to provide tenant sensitive, highly flexible, fully serviced accommodation, relieving businesses from property investment risks, property management problems, and day-to-day facilities management operations.

Innovations of this kind within the FM field must include the consideration of support service strategy and its contribution to core business operations at three levels of criticality. Firstly and most critically, there are the services that provide vital support to core business strategy and the key processes and operations of the organisation. Deficiencies here put the core lines of business of an organisation directly at risk. Second, there are routine but essential day to day services that provide quality support to an organisation's staff, customers and suppliers. In the past it has been customary for property professionals and senior management to undervalue the importance of the contribution that services such as security, cleaning, procurement, catering, etc., can make both to building and to organisational performance overall. Deficiencies here have negative effect on productivity and image of an organisation for its staff and customers alike. Third, there are outer shells of services that are rarely required but, when they are, they entail highly specialised knowledge that is not normally available within organisations. Architectural design and management consultancy skills are examples here.

FM is responsible for procuring and managing support services across this range, and for negotiating the specialist professional and consulting skills that may be required from time to time. So an important part of the FM business is to provide and manage these levels of support, maintaining an appropriate balance between long term property issues and the shorter term services support to day to day operations. Support services of this kind rarely receive the thorough attention that their position warrants. Chapter 5 addresses this area of neglect and sets out a logical framework for the positioning, appraisal and choice of business services strategy. It proposes a cross-sector platform for structuring support services arrangements in the future.

Chapter 6 sets out a comprehensive and highly ambitious approach to the integrated management of business assets. The Chapter provides a detailed account of the purpose, components and stages of Integrated Asset Management (IAM) comprising; the strategic asset plan, the asset acquisition plan, the asset management plan and the asset disposal plan. The IAM approach is in marked contrast with those set out in the earlier chapters along the business trail, perhaps indicating divergence between North American and

European positions.

Finally, chapter 7 looks at the alternative futures for the FM business itself. It provides a fundamental reappraisal of future performance issues for FM. A major change of direction is proposed in which FM would become more accountable to business, and on strictly business terms. The ambitious claims that have been made about FM's relevancy to business have been accepted in good faith but must need to be justified, with FM clearly demonstrating its own distinctive contribution to business advantage, business flexibility and agility. For this it must become more accountable to the core business, finding ways of measuring its own performance in contributing to business success. New performance indicators will need to be developed, that relate directly to strategic business objectives rather than levels of facility and service support alone. This need, for a fundamental repositioning of FM was the major conclusion of the conference debate along the business trail.

References

Economist Intelligence Unit (1999) *Vision 2010 - Forging tommorow's public - private partnership*, USA : EIU

Bates M. (1997) *Review of PFI (Public/Private Partnerships)*, London, HMSO.

H.M. Treasury *Private Opportunity, Public Benefit: Progressing the Private Finance Initiative*, London, November 1995.

H.M. Treasury Task Force (2000) *Value for Money Drivers in the PFI*, UK Treasury, London.

Montague A., Chief Executive, Treasury Taskforce (1999) The Private Finance Initiative - Directions for the Future, conference presentation *Futures in Property and Facility Management*, University College London.

3 Facility Management: opportunities, scope and impact

Oliver Jones, Citex Group Ltd

Overview

'Private Sector PFI' and its alternative 'Corporate PFI' are two new terms that have been increasingly seen in our industry during the last couple of years. Both reflect an interesting and seemingly wide recognition that there are potentially significant benefits to be gained in the private sector from studying and applying some of the ideas developed by the PFI/PPP programmes that have been pursued by the UK government during the 1990s. In particular it is believed that by using these techniques, the problem of the frequently encountered poor property fit faced by many businesses today can be solved by a more tailored and relevant accommodation solution.

For the government and public sector generally, the desire to avoid irregular capital commitments can be readily understood, as can the need to invest some of the capital generated in backlog maintenance and technical upgrades. Similarly the principle of only paying for services according to the performance achieved or quality delivered under such agreements is a major and welcome step forward. But is there more to the prospect than this?

For the private sector, these goals can equally be seen to be generators of real benefit for many. However, those major private sector businesses for which the current initiatives seem most applicable may reap far greater advantage.

The developments in this area may represent a fundamental change to many aspects of our industry - through outsourcing structures, property ownership and investment and the whole concept of accommodation. This paper sets out why this is the case.

The Opportunity

For many leading participants in the outsourcing industry today, the well understood tools of performance based or incentive contracting and improved techniques for operational management are now taken for granted as the starting point for services delivered. The top players - on both the in-house purchasing and industry supply side - are now however increasingly responding to core business needs. They are doing this by addressing a more strategic challenge using the concepts within PFI to construct service agreements that free core business capital tied up in facilities at the same time as reducing costs and increasing quality.

The opportunity has been presented by the UK economy and particularly the UK property market having strengthened in recent years (notwithstanding the slow down of late 1998). This has combined with public sector PFI deals having opened doors to the City and financial sector generally, leading to innovative facilities and property solutions being better understood and considered in both the public and private domains. Some fundamental questions arise:

- Is this the start of the new outsourcing generation where customers maintain an absolute core business focus and service partners extend their management to the ownership and operation of capital intensive assets such as property?
- And what does this mean for outsourcing managers tasked with exploiting these new financial led opportunities in terms of flexibility and the ways of supporting a business' increasingly dynamic (or volatile) strategies?

Whilst the answers are by no means generic or clear, the opportunities are certainly being crystallised.

The Focus of Outsourcing to Date

In its relatively short history in the UK, the outsourcing industry has been an exciting place to be as it has seen much change. Its focus has been refined from year to year following client-side challenges being creatively responded to by the leading supply-side players and the question has always been - what next? This question is just as valid today. In the main however, its actions and address have to date been necessarily introspective. The average quality of services and property management within many major businesses in the late 1980's left plenty of opportunities for improvement. Consequently there have been many areas demanding refinement and revolution. The in-house manager has therefore typically focused on getting the existing house in order - improving quality and reducing costs through partnerships, shared savings, mutuality with the market and so on.

We have seen these sequentially addressed through the emergence and cultivation of outsourcing protocols, the growth of the supply side and today the very exciting innovation sought by many blue chip multi-nationals and some public sector institutions, and increasingly offered by the top service providers. But where is the real opportunity to add further value going to come from? From which direction will the demand for continuous improvement be sourced?

Facilities Optimisation

Think about our world. As support services professionals, we often and quite appropriately focus on servicing the working environment that our clients or employers hold today. We may be directly involved and responsible for the property strategy and we certainly are involved in the optimisation of space within it - but we must not forget the basics. Every square metre of space we occupy costs us money - capital and revenue. Minimising space is one option, but how - by outsourcing and off-siting? In some cases, of course. Releasing capital however is another option - just as most of the industry has done with the formerly huge car fleets - so too, potentially, from property and other service assets.

Suppliers stepped up to the mark when the 'problem' of what to do with the existing internal teams created doubts for major corporates considering outsourcing. Substantial hosted joint ventures and external management buy outs followed and continue to be an attractive option. Today, however one of the biggest outsourcing challenges is driven by the disparity between the business world's needs and the property world's offering. Less and less fixed space will be required by major organisations in the future. The space that is required will also be very different from today - more flexible, and of a much higher specification. I believe this will become an increasingly important feature of major outsourcing deals in the future. Competitive advantage for suppliers will be derived from solving this major challenge for large corporates. The property angle is of particular current interest given PFI developments and I propose to explore this aspect in detail.

The Property Challenge

Much has been said in the past by organisational observers and writers such as Charles Handy and Warren Bennis on the changing work force. The forthcoming leisure era that was expected to be generated by the technological revolution - much heralded some ten plus years ago - has as we all now recognise, not happened. However, it is now recognised that the alternate options presented in more recent years by such commentators is very real as Handy suggested: half the people, paid twice as much producing three times the output.

This presents a real facilities problem. These 'super-producers' will need top quality environments to enable their productivity to be released. Few of today's commercial offices will make the transition without significant investment. These 'super-producers' will also demand space to suit them in all their working worlds - possibly various office locations and in their homes. The mid twentieth century legacy of traditional city centre corporate environments has a much more limited attraction in the longer term. In our new service oriented world, people will select from just two choices: to either work near their home or near their customer - or, very probably, both with their business colleagues - as business schedules demand.

So the challenge today is a real one for the outsourcing professional - how to distinguish the service offering by also guiding the organisations using current property portfolios towards a world where less and different property will be required. An easy option is to sit back and wait for the leases to end and plan the next twenty years accordingly, if the organisation can survive the competition being smarter. The tougher route is to try to read the market trends of today and use the dynamism of the outsourcing industry to deliver the flexible option sooner.

Some major organisations are in early and are trying to do just that already. They are exchanging their mixed bag of freehold and leasehold interests for much more flexible service occupancy agreements - in the same locations but with value being released in the form of cash, reduced running costs and real future space flexibility. They are actively moving out of direct property interests and disassociating themselves from the volatility and constraints that it brings.

The largest and best known example of these is the UK's Department of Social Security (DSS). Under an innovative outsourcing initiative code named PRIME, the DSS transferred all its 700 plus properties comprising around 18 million square feet and accommodating over 90,000 staff to a specialist consortium. The consortium paid an initial cash sum, and took over all assets (mainly freeholds) and liabilities (mainly leaseholds) and contracted to provide serviced accommodation on DSS defined flexible location terms for twenty years. The expectation and clear plan is that during this period, the total space occupied will fall, service standards will rise and costs will be reduced from day one.

In theory such moves sound ideal, but of course they are not feasible for all businesses. But as structural changes occur, the potential range of applicable organisations and portfolios will grow. Each internal manager should be monitoring this potential. Acknowledgement of the opportunities by the market will also start to increase as the PRIME deal and comparable private sector initiatives progress.

Property skills are vital, but the industry remains conservative and protectionist. Outsourcing management skills are equally important, but its leading expert organisations are still relatively young. The combination of the visionary initiators in both fields will be the key. Is this just a sale & leaseback concept? In some respects, yes, but in substance far from it, many new elements are now incorporated. The market is now moving on. All outsourcing professionals need to be aware of how and why and most importantly whether it will work for them.

To my mind all business support services providers, in-house or supplyside, have to examine their options on the big spend areas all the time. True market dynamics demand this. To date property has been a given, assumed fixed. This is, for some users and occupiers now no longer necessarily the case.

Property to Date

The institutional lease - typically characterised by a twenty five year duration, full repairing and maintenance requirement and upward only rent reviews every five years - has been the norm for the last thirty years. The terms were presented as non-negotiable, and standard in virtually all commercial deals. The interest of the building owner or landlord also rarely extended into positive occupancy support and service.

To my mind, as we face change today, I remember and draw loose parallels with the early reprographics contracts, which were equally one sided in service terms - those we look back on with astonishment just ten to fifteen years later. In property we must now ask - where is the equity and importantly the value for the tenant in such a deal?

The 1990s recession has knocked these base assumptions dramatically - few leases are signed today without more frequent tenant initiated breaks, and most importantly, twenty five years is no longer standard - around fifteen years is easily achieved on most new facilities. Why? Market demand. A early 1990s recession deeper than most can remember and consequently structural change potential within the industry has been forced.

When further changes such as the continuing development of outsourcing initiatives, UK investment market changes, public sector PFI products and the long awaited emergence of securitisation in the property market (albeit indirectly at this time through REITs) are added, a real opportunity beckons.

Is this just a property 'thing'? No, expert outsourcing initiation and implementation is key to the equation - as I shall illustrate. Service is everything in the more sophisticated markets and for those customers who are leading the field. For the outsourcing manager, this is yet another good reason to ride this particular industry wave. The returns may be good and the route exciting.

The Role of PFI in Change

The UK government's Private Finance Initiative (PFI) is important in this evolutionary tale. In changing the public sector's position on asset ownership, both the last and the current governments have been a major influencer in initiating change in the private property market.

The important aspect of the PFI drive is the recognition that to occupy property, a government department or agency does not need to own or even have an active interest in that property. Of far greater importance is certainty of expenditure, quality of service and value for money.

The resulting initiatives, mainly from the health sector but importantly including the DSS PRIME deal, have brought together different factions within the property, construction and outsourcing industries - financiers, property developers, constructors, facilities managers and service providers - to name just the primary groups.

Under PFI, separate payments for rent, rates, property management and support service packages, based around primary leasehold or freehold property interests are exchanged for a single Unitary Charge, for a serviced environment. The Unitary Charge being variable according to availability and quality of service received. As such the public sector client moves wholly away from buying assets and looking after them at risk, to using tax payers' money to purchase public services from the private sector - paying only for the service that is delivered in accordance with its defined requirements.

The consequence of the sheer scale of the PFI opportunity is a very simple one as far as outsourcing development is concerned. It has placed the activity at the core of major funding deals. Most importantly, all the leading City financial institutions that have looked at PFI, now recognise the vital role of the service provider. It is this that is creating the prospects for change in the private markets.

The New Property Prospects

As long ago as the 1997 British Property Federation (BPF) National Conference, Schroeder Chairman Win Bischoff commented on the next few years in the investment and property worlds, saying:

"The FRI (full repairing and insuring) lease will disappear, driven out as lessons from PFI spread to the private sector and bring about service occupancy contracts, with both the risks and rewards of ownership transferred to the owners".

This will be driven harder as securitisation becomes recognised as a way to fund PFI projects. The UK property industry in the late 1990s now appears by most measures to be back into a liquid and transactional form - after the fallow years which took their toll of some of the best known 1980s entrepreneurs. Not surprisingly, an increasing number have returned to the field in the last couple of years attracted in some cases by the potential presented by new PFI models.

The 'rules' governing the market and its operation continue to be under dynamic change; many ideas that have resisted the pressure to change within the last recession have been starting to cede to the even harder demands of a rising market. Traditional property and surveying firms and traditional FM operators will potentially face exposure in a more service oriented, value driven world as the leading outsourced service providers recognise the prospects - where is their added value proposition that everyone else has been pushed to provide? The innovators in the market are recognising that property, as with so many other businesses to day is seen increasingly in the eyes of the consumer and purchaser as a service not an asset.

Services can be differentiated to an immense degree - look at Virgin Atlantic. The core question is: who will lead the property market into this world?

Most major corporates would cite their need in property to be driven by their business, but the property market has historically never delivered. For so long, the institutional twenty five year lease was the proposed solution. But all markets and industries evolve when circumstances permit.

It is my view that the combination of pressures and opportunities today are such as to force new approaches from within the outsourcing sector.

The Way Forward to the Next Outsourcing Generation

If the PRIME initiative succeeds and delivers well, and there is every indication that it will, then it will become the rule for the major public sector organisations, and undoubtedly a guideline for the Corporate PFI potential faced by many major private sector businesses.

In all sectors, building users want space in the right but potentially changing locations on flexible terms reflecting their business needs. Current occupants could also benefit substantially from the potential release of capital through such transactions. A focus on core business and not property speculation should also be welcomed by many boards. The innovation however will draw from many related experiences gained in outsourcing in recent years.

From the perspective of where this takes the outsourcing industry, the answer lies in what will drive the building owners of tomorrow. Quality will be seen more readily in a more open and accessible market. The key in these cases will lie in having a focus on the occupiers' needs - the now established outsourcing principles of partnership being applied between investors, landlords and tenants. Full repairing and insuring leases will at one extreme make way for full business service occupancy contracts as in PRIME; still typically for ten to fifteen years, giving occupier stability but offering greater service scope.

At the other extreme, volatility beyond the core requirement and the need for customer or home adjacent facilities will be satisfied through totally flexible facilities from the serviced office sector and by increased teleworking - estimated by many to be on its way to around 20% of current total capacity. In both cases, the demand will lead the outsourcing industry participants of today to become integrated with or aligned to the other core skill providers from the financial and property worlds. New project based consortia will be the first step, radically integrated property investment and management vehicles will be the next. Full service environments can only thrive in such circumstances. Fully outsourced service packages may also then become the serviced environment norm - with purchasing power to the fore.

The Theme for the Future?

Increasingly high diversification by leading outsourced service providers being achieved within increasingly large externalised and financially leveraged portfolios.

4 Innovations in Facility Finance and Management

Michael Medlicott, Servus FM

This chapter explores the principal approach to financing property, the demand for flexible office accommodation and innovative responses to this demand. While the analysis is focused primarily on office accommodation it can be applied to other sectors. The Principal Finance approach has been pioneered in the UK by Nomura's Principal Finance Group (PFG). In the property sector PFG now combines Principal Finance with a long-term view of facilities management (FM) to provide 'a total property solution' for clients who need space but have no interest in the capital commitment, risks and management time inherent in owning property.

Evolution of Property Principal Finance

PFG's approach is about dealing quickly in a direct way and offering certainty of execution, simple documentation and direct decision-making, using Nomura's balance sheet strength to underwrite financing for large assets generating reliable cash-flow. Nomura may look to refinance transactions in the bank or bond markets but the transaction is not contingent on this refinance. Thus PFG is able to de-couple the terms of the transaction from market conditions and decision-making is not slowed by the need to reach agreement with a syndicate of co-investors or debt providers. Once the assets have been financed Nomura will work with existing management to develop strategy and empowers staff to drive the business forward. This approach can be applied to different asset classes: a key cash generating asset class is property.

Investment banks have always analysed the cash flow generated by property and other assets. However, the role of the investment bank has evolved and Nomura has fully developed the role to encompass principal investment together with the in-house capability to deliver a full property service. The most basic role for an investment bank is that of agent who will arrange finance for a property transaction on behalf of a client. The investment bank will receive a fee and will not have an equity stake in the profitability of the transaction. More sophisticated investment banks will arrange finance for clients using securitisation. This typically involves the income from the property being re-engineered to produce high quality cash flow which is sold to capital markets investors in the form of bonds. As the capital markets are among the largest and most efficient sources of finance available, a securitisation should produce the lowest cost finance. It is because of the ability of the capital markets to produce low cost, long term finance for high quality borrowers that governments raise debt by issuing bonds rather than borrowing from banks.

The next step is for an investment bank to act as principal in a property transaction rather than acting as agent. This means the investment bank finances the acquisition of the property, so taking more risk, which requires both an entrepreneurial approach and a significant balance sheet. However, if successful, the profits from acting as principal are much greater than the fees from agency business.

An investment bank acting as principal has an incentive to maximise its profits by developing property products to meet demand which is not currently satisfied. The demand for flexible, fully serviced office accommodation is an opportunity which can be met by simultaneously acting as principal and co-ordinating a consortium including FM providers. This was the approach taken by Goldman Sachs who established a consortium called Trillium, to buy the office estate of the Department of Social Security and provide in return flexible, serviced space (Project PRIME). This meets the needs of the client but the complexity of establishing and maintaining a consortium are considerable and a consortium falls short of the one-stop-shop sought by many clients. In order to meet the need in the market place, PFG established Servus to provide both Principal Finance and FM services without the need to form a consortium.

Principal Financier's View of Property

The Principal Finance approach to property is based on a view of property which differs markedly from the view of traditional institutional investors. The starting point for an investment bank analysing property is the cash flow and associated risks. An investment bank will seek to add value to that cash flow by understanding it, mitigating risks and then packaging the cash flow into attractive bonds. There are many different types of bond investor each with a particular profile of: attitude to risk; optimal investment term; currency preferences and interest rate expectations. Re-engineering cash flow into a package that will be attractive in the market place requires not only financial engineering skills but also a good understanding of the requirements of investors. An important area in which risk can be mitigated is to work with tenants and end-users to ensure the property is attractive and occupancy maximised. For example PFG works with tenants of pubs they have financed to provide them with competitively priced beer and other supplies and invests in the properties in order to enhance the underlying businesses and hence the rental income.

Traditional Institutional Investor's View of Property

Traditionally in the UK, equity in non-residential property has been funded by financial institutions that tend to hold long-term. Pension funds and life insurance companies have long-term obligations that may be inflation linked and hence they find long-term property income attractive. In particular, these institutions have been the driving forcing behind the traditional UK 20 year Fully Repairing and Insuring (FRI) lease with upward only rent reviews. Tenant default aside, this type of lease has provided investors with long-term, reliable income and limited risk. However, it does not give tenants the flexibility they increasingly need to respond to a dynamic business environment.

Institutional investors typically look upon property as:-

- A source of both rental income and capital gains as the value of the building appreciates.
- An asset class which diversifies their investment portfolio. Typically a major fund will have the bulk of its investment portfolio in equities and bonds. As movements in rentals and property values are often out of synchronisation with the stock market they help to diversify the overall portfolio and hence reduce its risk overall.
- An asset which requires some degree of administration. The FRI lease minimises risk for the investor but it still requires administration including rent collection, negotiation of rent reviews and finding new tenants when a building becomes vacant. Normally institutional investors minimise their administrative task and appoint agents to manage the property on their behalf.
- An illiquid lumpy asset which is expensive and time consuming to buy or sell. While shares in large companies can be traded at low cost at the push of a button, a property sale is normally an expensive and lengthy process that has to be individually negotiated.

The illiquidity of direct property investments can be mitigated by securitisation which turns property rental income into a liquid bond which can be easily and cheaply traded in a deep market. Illiquidity is one reason why an institutional investor may be reluctant to invest on the scale needed to have a significant direct stake in a large project such as Canary Wharf but is willing to invest in bonds backed by rental income from the project. In 1997, £555 million of bonds backed by rental income from Canary Wharf were issued. Not only do bonds allow investors to dynamically fine-tune their level of exposure to a particular property but they effectively relieve them of responsibility for property management. For these reasons institutional investors are substituting investment in property backed bonds for direct property investment.

Over the last twenty years there has been a consistent trend for UK life and pension funds to reduce the weighting of direct property in their investment portfolio. The percentage of their portfolio invested directly in property has fallen from 16% in 1977 to 6% in 1997. This shift by property investors away from direct property can be viewed as a re-balancing of portfolios to reflect changes in the risk reward profile of different asset classes, such as property, shares and bonds. Modern investment theory provides tools to determine the mix of different assets to produce a portfolio with optimal risk reward characteristics. Poor relative returns from direct property investment over the last two decades have driven the reduction in the size of the property component in an optimal portfolio.

What Tenants Want   The relatively poor returns from property investment may be linked with a failure by the UK property market to provide the full range of products wanted in the market place. While many property users have a core portfolio they wish to retain over the long term (e.g. corporate headquarters), they are also likely to have accommodation needs which will fluctuate with the state of the market, operational changes (e.g. such as shifting the back office from London to Cardiff or Calcutta) etc. The demand for flexible space is not well served in the UK and the property market has been slow to grasp an opportunity to offer a product that will command a premium price.

Demand from the Government

In most countries the government is the largest single occupier of office accommodation and uses a mix of freehold and leasehold property. As the major occupier, any significant shift in the government's position could have a big impact. In the UK, the government has signalled a major shift in its property strategy that is being implemented through a series of Public Private Partnerships (PPP) transactions (these were originally called the Private Finance Initiative (PFI)). The review of PFI by Sir Malcolm Bates in June 1997 concluded that:

> "Government bodies are moving from being owners and operators of assets into becoming intelligent purchasers of long-term services."

As far as property is concerned, the Bates review signalled that in principle the government is moving from owning or leasing offices to buying property services. The government is no longer directly interested in property but instead wants flexible space in which to carry on its core activities effectively. The sale of government offices in this way is entirely consistent with the 'third way' and its pragmatic focus on managing to achieve solutions that work by combining the best of the public and private sectors. This important policy development is a logical extension of the now widely accepted argument that government does not need to own productive assets such as British Telecom (BT) and British Airways (BA). By selling these assets the government ensured that they are operated more efficiently and government is freed to focus on its core areas of responsibility.

Demand from the Private Sector

The private sector is following the government in demanding serviced space procured so as to cut costs and enhance the efficiency of core activities. The drivers of this demand include:-

- The release of capital tied up in low yielding property assets to be redirected more productively to core activities. For example, there are clear gains for a pharmaceutical company from selling property yielding circa 8% and investing the proceeds in research and development of new drugs where the expected rate of return is double or treble.
- A need to have an office estate which changes in response to developing business needs rather than constraining flexibility. Competitive pressures place a premium on the ability to change the type of accommodation to keep pace with changes in office technology and working practices, to move offices and to expand or contract the estate.

- The elimination of risk exposure to the property market and the outsourcing of estate management which demonstrates a clear focus on core activities. This will be well received by shareholders and financial analysts who appreciate the value added which comes from a clear specialisation.
- Cuts in the cost of using property as FM costs are driven down, office space is used more efficiently and development opportunities are fully and rapidly exploited.

The Nomura Property and FM Offering

Nomura has continued to innovate in the property arena by establishing Servus to meet the demand for cost-effective, flexible serviced space. This has required a dedication to delivering customer focused services at the lowest possible cost. Servus' customer focus is most apparent in young, highly educated and motivated professionals who are fully trained to take ownership of problems and solve them. They are given all of the required authority and equipment to resolve problems to the client's satisfaction.

The close strategic and operational integration of property strategy with FM helps to reduce costs. A key aspect of the life-cycle maintenance approach is investing to save by upgrading plant and equipment to extend equipment life or to reduce costs. FM is also enhanced by ensuring it is informed by plans for the building: for example adjusting refurbishment or renewal when a building is not required in the long run. In addition, designing operational FM considerations into new buildings or major refurbishments will reduce life-cycle costs and improve service levels.

The potential for this integrated approach to property and FM to deliver value and assist clients to achieve their objectives is greatest with large estates that are geographically diverse and freehold rich. Economies of scale in procurement and management mean that this approach is best suited to large estates. Geographic diversity reduces the cost of providing clients with flexibility and an exposure to a number of different regional property markets mitigates the property risk. Freeholds have inherently greater scope for redevelopment and so an entrepreneurial approach can realise more value from freeholds than from leaseholds.

Conclusion

There is a demand in the market place from major organisations to escape from the cost of owning property and facilities, to release cash for core activities and to be supported by serviced accommodation. Servus has been established specifically to meet this demand with all of the expertise needed in-house to do so successfully. An innovative approach to finance and FM is employed to provide a value added total property solution in a long-term strategic alliance with public and private clients.

References

Bates M. (1997) *Review of PFI (Public/Private Partnerships)*, HMSO, London.

NAO (1999a) *The PRIME Project: the transfer of the Department of Social Security estate to the private sector*, Department of Social Security, HMSO. London.

NAO (1999b) *Examining the value for money of deals under the Private Finance Initiative*, HMSO, London.

5

The Business of Support Services Strategy

Stephen Bennett, Facility Management Consultant

Support services strategy is a neglected area of concern for many organisations. This is surprising given that support services are a recurrent issue of strategic concern for the operation of all organisations. There is a need to move beyond reliance on best practice approaches and to consider strategic frameworks for the logistical support of any organisation and its business.

Support Services

There is a range of views of support services in the facility management field. Three particular views are recurrent. These are shown in Figure 9. Firstly, support services may be considered as a group of dis-aggregated packages. This view, which is understandable given the disparate disciplines now involved in facility management, sees support services as one of a number of areas requiring management attention. The second view considers the field in terms of a physical infrastructure envelope, or more narrowly a property envelope, with support services contained within it. Such a perspective is reasonable given the influence of physical infrastructure in the history of facility management, and is particularly relevant where the physical asset issue is the main concern, such as may be the case in some areas of investment. Thirdly, the field may itself be considered as a support service envelope, with property seen as an example of a package within it. This view represents a shift from product-based to service-based strategy. Whichever of these three views is taken, support services feature as a significant element. However, it is the perspective of a support service envelope that seems to have significant potential as a platform for innovation.

Concerns in Support Services Strategy

Support services strategy lacks generic frameworks. The critical differences between case specific and generic investigation is a matter of general concern in the facility management field (Nutt, 1998) and is discussed in some detail in Chapter 21. It is a source of particular difficulty in support services strategy. The relationships between generic and case specific work in support services strategy are shown in Figure 10.

There has been a tendency for support services strategy to be dealt with on a case-by-case basis, which poses serious doubts about generalisability. The case-by-case approach may be sufficient for maintaining a support operation in its current form, but is likely to run into problems when strategic change is required. The sector-based approach is a reasonable attempt to move beyond the case-based approach. This view is rooted in the idea that organisations in similar business sectors face similar logistical support problems. The sector approach manifests itself as a form of specialisation in practice, by both clients and suppliers, and to some extent among academics. However, the sector approach means that serious questions remain concerning fundamental differences that may exist between organisations in the same sector or in a particular organisation over time, and significant similarities that may exist between organisations in different sectors. This suggests that generic approaches, rather than those that are case-based or sector-based, are worth consideration.

Figure 9   Alternative views of facility management.

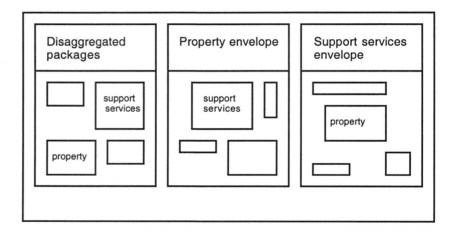

Figure 10   Generic and Case Specific Investigations.

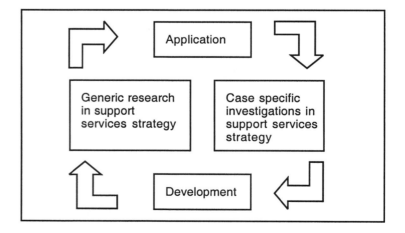

Addressing one support service at a time and one strategic issue at a time compounds the problems of case-based, or even sector-based approaches.

This results in a serious lack of coherence in the area. Polarised debates arise on individual services from catering to construction on such issues as outsourcing and insourcing, service integration and differentiation, and whether to benchmark or not. Such approaches lead some to advocate a particular position for a support service as being so-called best practice. While the concept of best practice may be helpful in highlighting some of the issues of concern in support services, it is fundamentally flawed in the area of support service strategy and so does not provide a secure platform. There are two reasons that independently undermine the notion of best practice in support service strategy, which if taken together are damning. Firstly, the strategic position of support services needs to take into account the contingencies that exist for a particular organisation over time. These contingencies may be narrowly or widely divergent for the organisation over a period and between organisations at a point in time. These issues must be taken into account. Secondly, the notion that there is some optimum is often impractical given the complexities and uncertainties inherent in strategic decision. More modest and circumspect approaches are required, such as recognising that satisficing might be a more realistic target.

Indeed more specific weaknesses of current approaches to support services strategy based on best practice can be identified. Chief among these is that the dynamic relationship between decision and position factors in context is ignored. This means three crucial features are not addressed adequately. Firstly, critical issues in support services strategy tend to be obscured, in favour of over-focusing on fashionable topics. Secondly, the relationships between these issues are seldom explored fully. Thirdly, the dynamic nature of the issues of concern tends to be disregarded. The result of these failings is that important strategic dilemmas in support services are not faced, still less resolved

Table 1   Support Services Orientation.

Past	Future
Area of management neglect	Focus for management attention
Property envelope perspective	Support services envelope perspective
Case or sector based approaches	Generic approaches
Best practice reliant	Appropriate practice based

The implications of this are far reaching, because the strategic positioning debate is robbed of its proper organisation specific context, and management is blind to the vital path dependent nature of the strategic decision process.

What is required is a contingency approach that provides a strategic platform for appropriate decision concerning support service positions. Such a platform must address the critical issues, the relationships within and between strategic decision and positions, and the dynamic characteristics of the issues of concern. Issues relating to the past and potential future of support services strategy are summarised in Table 1.

Requirements for a Cross-sector Contingency Approach

The proposed perspective requires an approach that is contingency based, generic, and grounded in a support services orientation to facility management. A contingency approach requires that the effects of a range of recurrent circumstances and events be considered in a dynamic context. There are three main components needed in order to establish a contingency approach to support services strategy. The first two parts involve the identification of important factors in support services positions and in decisions, and the third part is to develop a framework to relate these two together. These critical aspects provide not only a basis for considering strategic positioning, but also a framework for classification of support service strategy that relies on significant similarity and dissimilarity that is independent of any sector or particular organisation. The proposed framework for support services strategy is founded on considering support services as operations, decisions as processes and achieving a strategic dialogue between operating positions and decisions. Each of these three components is now addressed in turn.

Strategic Positions

Support services can usefully be considered from an operations perspective. An organisation can be considered as a dynamic matrix of operations consisting of support and primary operations. This is shown in the Figure 11.

Figure 11   Operations Matrix.

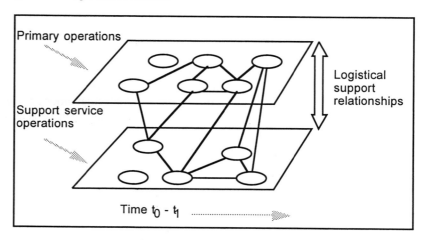

Figure 11 shows two planes indicating the support service and primary operations of an organisation. Each plane shows a group of operations that are linked together in various ways. Crucially, the relationship between these two sets of operations is a logistical support relationship through which the primary operations are served by the support service operations. Various configurations of these links are shown. The period t0-t1 indicates that the operations and the linkages between them are dynamic. This matrix provides a basis for considering operation arrangements and associated management arrangements, which can be described briefly.

Operation arrangements can provide a starting point for considering strategic positions. Operation inputs consist of mixes of people, physical and informational resources, which can be acquired through financial resources. The generic processes that characterise these operations relate to the key ways in which they add or subtract value, which include changing or avoiding and reversing change in form, location and ownership, and management of a recipient. The recipient may be people, physical entities and materials, or information (Morris and Johnston, 1986). This allows operation outputs to be considered as products. However, the defining factor in this area of concern is the logistical support relationship between support service operations and primary operations. This can be conceptualised as having dimensions such as specificity, criticality and frequency. Each of these dimensions can be considered further; for example, the degree of criticality may vary by organisation, operation, and time. Recent and historical military conflicts have provided extreme examples where the importance of logistical support can be readily appreciated. The humanitarian crisis in the Balkans in 1999 (Franchetti et al, 1999) graphically illustrated the impact of logistical support on the refugees, covering operations from the provision of food, shelter and medical help through to record keeping, counselling and child care. The importance of logistical support has been long understood in military organisations (Keegan, 1994). However, this does not mean that such organisations are always successful in this area of management. For example, in the winter of 1854-55 during the Crimean War, under the British army's troop welfare regime, Raglan's army suffered a thirty-five percent decline in active strength despite there being no combat (Dixon, 1994).

Operation arrangements need to be managed by people in organisations. One starting point to consider these management arrangements is the transaction relationships between providers and customers. These relationships have various dimensions, although chief among these is the nature of the perceived exchange, which at the extremes can be seen as transactional or relational. This approach allows the pure forms of management arrangements to be classified as markets, clans or bureaucracies (Ouchi, 1980), in both inter-organisational and intra-organisational terms. Each type of management arrangement has a characteristic way of planning and controlling operation arrangements with implications for alternative methods of performance measurement.

Operation arrangements and management arrangements can be linked together through a framework incorporating networks of supply, interface and demand. This gives a way of considering the packaging of support and primary operations with providers and customers. It also gives a way of linking logistical support relationships with transaction relationships. Although classic transaction cost economics provide useful insights (Williamson, 1981), it is necessary to go beyond this, not only to consider value, but to move towards a richer multi-dimensional view of negative and positive impacts. Usually, a demand driven approach is adopted, where the needs and wants of customers are the main concern, and supply is changed accordingly. This puts the onus on managing customer expectation at the interface whether this through some form of help desk operation or particular people. However, the converse view of a supply driven approach can also be considered, and can be at least part of the solution (Moody, 1992).

Strategic Decisions   Given that consideration of the support service positions is required in a dynamic context, it suggests that decision making be process oriented rather than static and calculation based. However, there is more than one mode in which strategic position decisions for support service operations can be made, each with important differences in emphasis. Some decision modes that occur in support service strategy can be outlined together with the broad decision environments in which they might be more or less appropriate.

Two types of decision mode are recurrent. Firstly, there is the process mode of which three principal kinds are synoptic, incremental and disjointed modes. Advocates of each of these modes for use in strategic decisions can be found (Andrews (1980), Quinn (1978) and Lindblom (1990) respectively) with most strategic management literature weighted towards the synoptic approach. There are important differences in the approach taken to decision in each of these mode. The most significant of these is the view taken concerning rationality. Management takes a relatively optimistic perspective in the synoptic mode, a relatively pessimistic view in the disjointed mode, and an intermediate approach in the incremental mode. A second type of decision mode emphasises either the current position, changes in positions or future positions.

One approach with encouraging potential for consideration of the contingencies in which the modes of decision might occur is through the lens of the generic features of the decision situation, (Cray et al, 1991). On the first dimension, the decision situation contingencies can be sub-divided from the fundamental features of complexity, such as number, variety and uncertainty. Generally, the more complex the decision situation the more likely it is that the mode of decision shifts from approximations to synoptic through logical and on to disjointed. A large number of stakeholders involved in a decision, all with significant influence, makes a decision more complex. A high degree of threat involved in a decision increases uncertainty. Where large numbers of stakeholders are present and there is also a high degree of threat then the decision situation is highly complex and uncertain, and so it is more likely to proceed in a mode closer to the disjointed end of the scale than the synoptic end. Similarly, on the second dimension the decision situation goal can be considered, since whether the problem faced is broadly negative, neutral or positive is important (Mintzberg et al, 1976). The general aims of moving away from a current position, moving towards a future position, or moving about to explore a more or less neutral zone, are fundamentally different paths

Figure 12 Decision Complexity and Process.

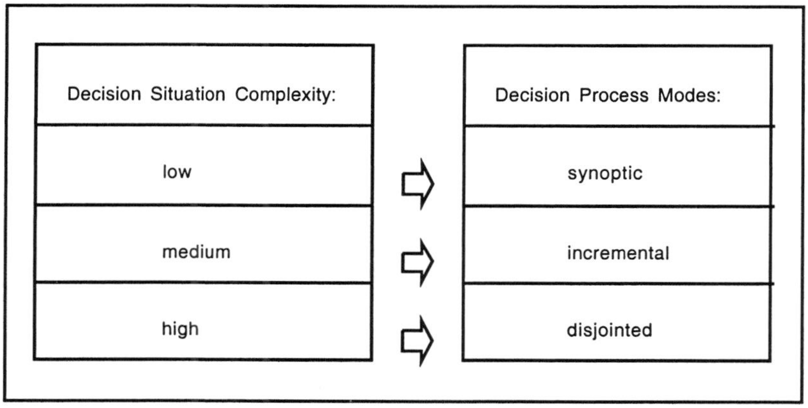

One of the links between the nature of the decision situation and decision mode is shown in Figure 12. In practice, a mix of decision modes is found in support service positioning. One effect of this is the potential for mis-match between the decision modes of clients and consultants. This may have benefits in terms of encouraging challenging debates, but runs the risk that there may be serious breakdowns in communication.

Relationship Between Strategic Decisions and Strategic Positions

Ultimately an approach that links decisions and positions in support service strategy is required. The nature of the support service operations and their dimensions may in part affect the decision situation and so influence the decision mode. The reduction of a supplier base from a large number of suppliers with little power to a small number of suppliers with significant power will potentially create a different decision situation for the subsequent decision. The decision mode may affect the position of the support service operation. A synoptic approach may lead to an integrated solution with positive co-ordination of services, such as prime contracting, while a disjointed view may emphasise negative co-ordination that seeks to avoid serious clashes between suppliers whether these are internal or external. Decisions, positions and their interrelationships are incorporated in a framework in Figure 13.

Figure 13 encapsulates the main generic strategic dilemmas previously identified that other approaches fail to address. The major recurrent issues in support services strategy of decisions and positions are indicated in outline, together with the general form of relationships within and between them, all in a dynamic context. The support service operations under consideration have various operating dimensions. These dimensions are the critical features of the strategic position when considering an organisation as a matrix of operations that need to be managed. Within decision are shown decision modes, which are the principal ways in which the decision process operates. Support service operations are shown together with their contingencies in the form of associated operating environments. Decision modes are shown with their contingencies in the form of associated decision situations. The two-way relationship between decision and positions indicates that these two recurrent issues are connected in an important way. Two-way relationships are also shown between operating environments and support service operations and between decision situations and decision modes. The emphasis is from operating environment and decision situation to support service operations and decision modes respectively. This is to indicate that usually the effect of these contingencies on support service operations and decision modes is the main focus of attention, but that the converse relationship is also worthy of consideration. The dynamic context of positioning for both decision and positions is denoted by a time period t0-t1.

Figure 13   Positioning Strategy for Support Services.

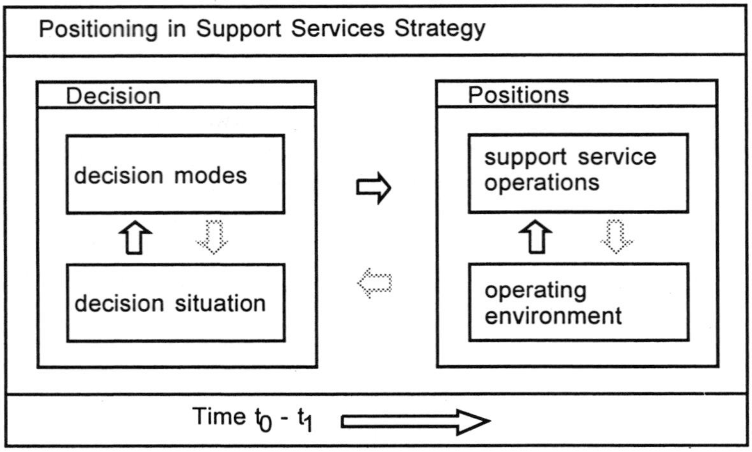

```
┌─────────────────────────────────────────────────────────────┐
│  Positioning in Support Services Strategy                     │
│  ┌──────────────────────────┐      ┌──────────────────────┐  │
│  │ Decision                 │      │ Positions            │  │
│  │  ┌────────────────────┐  │      │  ┌────────────────┐  │  │
│  │  │ decision modes     │  │  ⇒   │  │ support service│  │  │
│  │  │                    │  │      │  │ operations     │  │  │
│  │  └────────────────────┘  │      │  └────────────────┘  │  │
│  │      ⇧      ⇩            │      │      ⇧      ⇩        │  │
│  │  ┌────────────────────┐  │      │  ┌────────────────┐  │  │
│  │  │ decision situation │  │  ⇐   │  │ operating      │  │  │
│  │  │                    │  │      │  │ environment    │  │  │
│  │  └────────────────────┘  │      │  └────────────────┘  │  │
│  └──────────────────────────┘      └──────────────────────┘  │
│              Time $t_0 - t_1$   ⇨                              │
└─────────────────────────────────────────────────────────────┘
```

Implications of the Approach

Implications of the interrelated decision and position contingency framework can be considered. The overall implication of the approach for organisations is that there is no single best strategic position for support service operations or best way for their strategic decisions. This is a rich area for discussion, but some of the more deterministic implications can be outlined.

The mode in which a strategic decision is made affects the support service position that is selected. An implication of this is that the managers of organisations with similar support services and operating environments, but different decision situations, may formulate different operating strategies. The management of a manufacturing company under threat of closure may logically formulate a different strategy to that of a similar company where the management has no such concerns. The operating environment affects the support service operations. Consequently, in organisations with similar support services but different operating environments, management may formulate different strategic positions. The criticality of security provision in one financial services business to another may be significantly different, but may have similarities with that in a government department. Strategic decisions are processes and support services are dynamic operations. Consequently, the management in organisations with similar support services and strategic positions at a point in time may find that those strategies diverge over time. The management of two NHS Trusts may find that their strategies may logically become increasingly different over time with changes in operating environments and/or decision modes.

It is important to focus on the similarity and dissimilarity between organisations in decisions and positions. This enables what is appropriate in support services strategy to be identified independent of dogma or fad.

Applications of a Cross-sector Contingency Approach

The approach has three main areas of application. Firstly, the issues of concern in support services strategy can be clearly identified and described, which provides a basis for focused debate. Secondly, retrospective application to cases allows past decisions in support services strategy to be reviewed. This will have a valuable role in helping to inform future work, although may be argued to have practical limitations, as it is a 'rear view mirror' application. Consequently, the third and most valuable application is prospective in real-time, as it enables the method to be used as a guide to support decisions regarding future support services strategies. This provides a basis to explore areas of promise, with the result that appropriate and potentially innovative strategies can be formulated. The approach is not role-specific, and may be used by suppliers, intermediaries or clients acting in an advisory capacity. To realise the potential of this overall framework it needs to be converted into a practical expert method that comprises a set of techniques and tools, and this is the subject of current research (Bennett, 2000).

More generally a support services orientation to facility management, or merely serious consideration as to their importance, has significance for facility management policy in key areas. Practice, education and research are three such aspects in this region of applied management, which are related and are arguably all part of the same multi-disciplinary and inter-disciplinary field (Kubr, 1996). In practice the support service orientation will continue to provide a platform for the development of sophisticated suppliers, consultants and clients, as Jones (above) says "Service is everything". In terms of the professional bodies there may be increasing tension between those that are rooted in physical infrastructure and property investment and the service-led approach of individuals and organisations in practice. A modest expectation in education is that managerial and technical frameworks in support services strategy, with cross-sector applicability, will be developed to address support services at an appropriate level of significance in facility management programmes, particularly at masters level. A more radical outcome may be that a support service perspective will provide the basis for a fundamentally different orientation to facility management courses at strategic level run by state and "corporate" universities, with physical infrastructure being treated as one important aspect. The kind of framework discussed here may provide the foundation for research into contingency approaches to strategy in support service operations. Its generic nature means that there is potential for it to be used to study any organisation regardless of type. The connection between generic research and case specific investigation means that collaborative work is beneficial to both practically minded academics and academically minded practitioners.

In summary, a support services orientation that is based on a generic, rather than a sector, perspective is required in order to move beyond best practice approaches. In support services strategy, a framework that incorporates a contingency approach to decisions and positions in a dynamic context provides such a basis. The orientation and framework have implications for facility management policy, organisations and the development of a method based on secure practical theory.

References

Andrews K. R. (1980) Director's responsibility for corporate strategy, *Harvard Business Review*, November-December, pp. 30-44.

Bennett S. R. (2000) draft PhD thesis, University College London.

Cray D.et al. (1991) Explaining decision processes, *Journal of Management Studies*, Vol. 28, No. 3, pp. 227-251.

Dixon, N. (1994) *On the psychology of military incompetence*, Pimlico, London.

Franchetti M. et al. (1999) War in Europe Focus Special, *The Sunday Times*, 11 April, pp. 13-14.

Keegan J. (1994) *A history of warfare*, Pimlico, London.

Kubr M. (ed) (1996) *Management consulting: a guide to the profession*, 3rd edition, International Labour Office, Geneva.

Lindblom C. E. (1990) The science of 'muddling through', In: *Organization theory: selected readings*, (Ed. by D. S. Pugh), pp. 278-294. Penguin Group, London.

Mintzberg H., Raisinghani D. and Theoret A. (1976) The structure of "unstructured" decision processes, *Administrative Science Quarterly*, vol, 21, pp. 246-275.

Moody P. E. (1992) Customer supplier integration: why being an excellent customer counts, *Business Horizons*, July-August, pp. 52-57.

Morris B. and Johnston R. (1986) Dealing with inherent variation: the difference between manufacturing and service? *International Journal of Operations and Production Management*, Vol. 7, No. 4, pp. 13-22.

Nutt B. (1998) Moving targets, *Facilities Management World*, No. 11, July/August, pp. 25-28, 30.

Ouchi W. G. (1980) Markets, bureaucracies and clans, *Administrative Science Quarterly*, Vol. 25, pp. 129-141.

Quinn J. B. (1978) Strategic change: 'logical incrementalism', *Sloan Management Review*, Fall, pp. 7-21.

Williamson O. E. (1981) The economics of organizations: the transaction cost approach, *American Journal of Sociology*, Vol. 87, pp. 548-577.

6 Integrated Asset Management

Edmond Rondeau, International Development Research Council (IDRC)
Tony Stack, FM Strategies,
Elizabeth Fox, Environment Canada

In a business environment of mergers, corporate take-overs and buyouts, the surviving organisation must give the highest priority to ensure its long term financial viability. This requires a redirection away from traditional capital investment practices and a strengthening of infrastructure management regimes; that is Integrated Asset Management (IAM). This chapter reviews the concept, development and trends of IAM, looking at the cost and business issues that IAM addresses.

The adoption and implementation of Integrated Asset Management throughout the organisation can be monitored as part of the overall process of capital asset planning and implementation. This Integrated Asset Management planning process can also be integrated with the Corporate Strategic Planning cycle. Integrated Asset Management may be defined as the sum of all those activities that result in appropriate infrastructure for the cost-efficient delivery of service. These activities have the following strands:

- The identification of the need for an asset.
- Providing the asset, including its renovation.
- The operation of the asset, including its maintenance.
- The disposal of the asset from an organisation's portfolio, through its effective removal.

In this context, asset requirements are driven by business and service needs and should not be viewed as a need in themselves. This ensures that scarce capital resources are properly allocated and managed to maximise the return on investment.

The Concept and Application of IAM

Integrated asset management is a concept which organisations use to optimise and integrate the number, type, size, location, initial capital costs and on-going maintenance expenses, for both infrastructure that is owned and infrastructure that is leased. The concept also identifies the quality of facilities an organisation chooses to provide for the people whom it employees, the cultural values the organisation seeks to promote, including quality of life within the community and environment, which must be able to support and sustain all facets of the organisation, now and in the foreseeable future.

Integrated Asset Management is a tool to match corporate infrastructure resource planning and investment with corporate service delivery objectives and criteria that include:

- An asset has an enduring value.
- Assets should exist to support service delivery strategy.
- Assets are only one input in the corporate planning process.
- All asset must have a service potential.
- Assets can be financial, physical or intangible.
- Assets can be current or non-current.
- All assets need to distinguish between their physical and functional life expectancy.

International Trends

The impact of trends in the management of infrastructure supporting service delivery to the community is forcing organisations to adopt innovative and integrated approaches to planning and decision making. The concept and use of IAM has both national and international applications. It has become a logical process for looking at:

- The globalization of markets.
- Pressure on cost of service and competition.
- Acquisitions, mergers and take-overs.
- Common products and profit margins.

Integrated Asset Management as a concept has been practised to some degree by centralised organisations for many years. Few have embraced the concept fully because of the strong centralised research and decision concepts which must be delivered over a long time period often three to ten years. The concept requires that organisations follow through with quality research, an objective, not subjective, process and a full program of option development before decisions are made. From a historical viewpoint the process requires that within the organisation there:

- Is a demand for efficiency and effectiveness.
- Is a perceived 'piecemeal' approach which is not working well and needs to be changed.
- Is a threat to organisation's perceived and actual wealth.
- Is a management framework which is 'robust' and is willing to make a financial, cultural and political commitment to IAM.
- Are viable internal processes which are compared to current and proposed outcomes.
- Is a current lack of an overall co-ordinating body at the senior management leadership level for asset management.
- Is a high cost of asset creation.
- Are identified and validated costs for maintaining and staffing the organisation.
- Are ageing assets with strategic maintenance program requirements.
- Are changing expectations from employees, especially via technologies.

- Are financial risks to the organisation.
- Are fixed assets, capital fund requirements, and service delivery needs.
- Are better infrastructure investment with community and industry service needs.
- Are alternative ways of delivering services that do not require creation of infrastructure.
- Are planning tools and decision tools.
- Is a performance monitoring program.
- Is a best practice program for energy and environmental management; real estate and facility management; human resources and personnel services; information technology; financial management; asset management; heritage assets; total cost planning and community infrastructure development. Both public and private sectors are using the concept in a number of countries, while in other countries the concept still requires additional research and locally defined benefits through case investigations before being used on a regular basis.

Private Sector Industry Trends

A number of organisations in north America have embraced IAM. This has required senior management at the highest levels to have the vision and determination to understand and apply the concepts in a proactive course of action that requires a multi-year commitment to achieve success. The facility professional must not only deal with the realities of day-to-day work, costs and the politics of IAM but must be willing and able to address IAM requirements to:

- Be customer and business focused.
- Be an on-going internal programs to reduce assets, costs, and head count.
- Looking at business infrastructure, services, and costs from an integrated and portfolio perspective.
- Outsourcing non-core services as an integral part of the IAM program.
- Establish and use financial and performance measurement and benchmarking programs.
- Team with internal and external service providers.
- Developing Partnering, Alliances, etc. as a strategic method of doing facility management business.
- Use a Business Case Analysis (BCA) approach to identify current business methods, identifying risks to the organisation and to develop risks assessments which lead to recommendations for changing the way facility and organisational business is conducted.
- Identify ways to identify Risk Sharing and the realities of Fee at Risk contracts with service providers.
- Develop and use Service Level Agreements with facility customers.
- Use a Balanced Scorecard with customers to assess and quantify performance.
- Use Off-Balance Sheet Financing when and where appropriate especially in a sale-lease-back environment.
- Use Quality and Performance Measurements to establish a baseline of ser-

vice and to establish progress over time toward improvements in facility quality and performance, as shown in Figure 14.

Figure 14   Quality Measurement Processes.

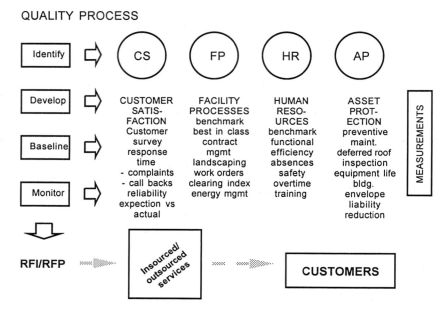

QUALITY PROCESS

IAM Principles and Processes

Integrated Asset Management is a tool to better match infrastructure planning and investment with community and industry service needs with the following considerations:

- Review of the organisation's broader economic strategy.
- Co-ordinating divisional and/or business unit budgetary process.
- Organisation wide, whole of life asset management in the context of one single co-ordinated program delivery framework.
- Best Practice management including the development of a detailed Service Management Process (see Figure 15).
- Develop a centralised co-ordination and management process for real estate and facility strategy development and implementation to ensure that:
  - Real savings are provided to the organisation regardless of the strategy is implemented;
  - The IAM approach maximises organisations purchasing power;
  - There is a realistic and verifiable return on investment;
  - The strategy provides improved performance standards for facility accommodation, management, and disposal.
- Sustainable Development must be incorporated into the Integrated Asset Management policy and program where leadership and services supports an organisation's (public and private sector) commitment to maintaining

its stewardship role in regard to long-term economic development, social responsibility and environmental sustainability.

IAM Objectives

An organisation that chooses to use IAM does so for competitive business and sustainable growth reasons and as a process or solution to address those competitive pressures. IAM objectives are strategic and are designed to:

- Achieve the best possible match of assets and service delivery strategies.
- Reduce demand for new assets by adopting 'non-asset solutions'.
- Maximise the service potential of existing assets.
- Lower the overall cost of owning assets through life cycle costing techniques.
- Ensure 'bottom line' focus by establishing clear accountability and responsibility.

Figure 15 Service Management Processes.

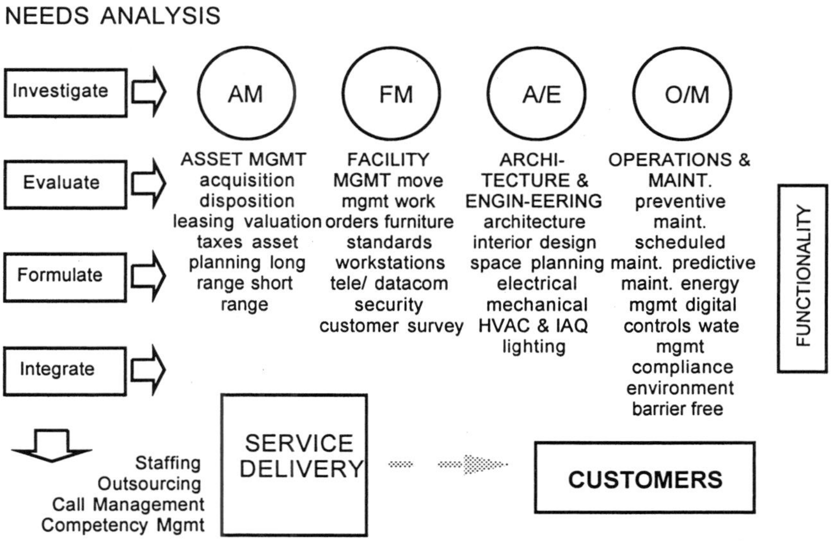

NEEDS ANALYSIS

The Master Plan

Within Integrated Asset Management there is a master planning process which is shown in Figures 16 and17. It includes:

- Identifying the problem.
- Describing the problem and implications if the problem is not remedied.
- Research & benchmarks.
- Desired outcomes and objectives.
- Individual strategies and integration of selected strategies.
- The net cost of implementation including capital, operational, personnel and associated program costs.
- The net savings to organisation both in actual and intangible Dollars, time and competitive advantage.
- Selling the concept.

The first step in developing the Integrated Asset Management Master Plan requires the identification of IAM problems that could include:

- Higher than industry average occupancy costs.
- Average maintenance expenditure on owned assets outstripping organisational funding.
- Divisions or business units that have sole control of property acquisition and disposition.
- Diminished organisational purchasing power.
- Ageing facility/property stock with reducing flexibility.
- High ownership ratio of facilities and real estate.
- Increasing churn costs as a result of organisational change costing millions per year.

The Master Plan includes four key components: the Strategic Plan, the Acquisition Plan, the Management Plan and the Disposition Plan.

Figure 16   Integration of Strategies at Asset Level.

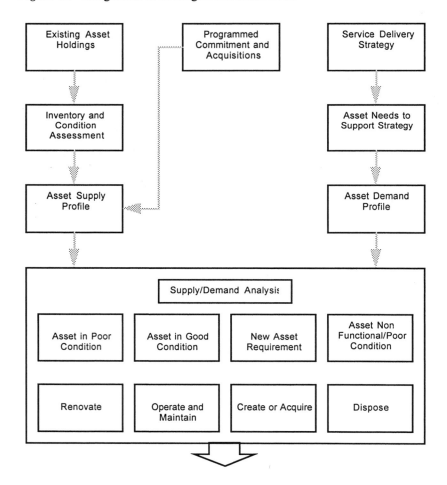

The Strategic Plan

The Integrated Asset Management strategic plan is designed to develop an integration of strategies at the corporate level of the organisation as shown in Figures 16 and 17. To be effective the strategic plan must:

- Access asset functions against and matched with service delivery strategies, including the planning time frame against the corporate planning horizon.
- Strategically incorporate capital, personnel, one time expenses, and recurrent costs which are linked to the financial management strategy.
- Result in an integration of strategies into business plans to optimise the service potential of new and existing assets.

The Acquisition Plan

This plan requires a through research program with alternatives and options identified which includes:
- Plans and details for the rationale acquisition/ replacement of sites/facilities.
- Facts which establish that existing assets are fully utilised and meet service delivery requirements.
- Consideration of 'non-asset' solutions.
- All express and implied costs in Life Cycle Analysis.
- A review of all Infrastructure Management, Client Relationship issues and the issues.
- The result should be an efficient and effective acquisition framework which will reduce the demand for new assets, funds, and enhance the organisations business programs.

Figure 17   Integration of Strategies at Capital Level.

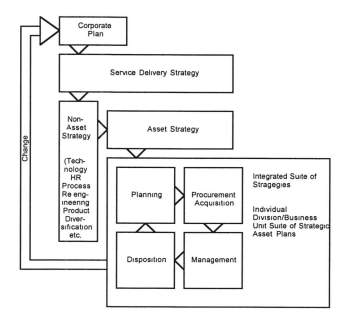

The Management Plan   The Management Plan provides reliable, relevant and timely data with which to make informed asset management decisions and should include:

- Policies and procedures.
- Asset registers and accommodation databases.
- Internal/actual rent charge back systems.
- Infrastructure Management including integrating all corporate facility/asset services such as: Portfolio Management, Corporate Real Estate, Facility Management Services, Human Resources, Information Technology, Telecommunications, Purchasing, etc.
- Result an effective internal control system provides the framework within which improvements to assets are effected.

The Disposition Plan   Embarking on an IAM program will identify assets which should be considered for disposal. Legal, financial, personnel, community, technology, and cultural issues must be considered in the decision making process to determine:

- Which assets are utilised, are under-performing assets, and have been identified as part of a systematic review.
- Reasons for under-performance and corrective asset action can be taken.
- An analysis of disposal methods with regard to potential market, timing and ability to maintain service delivery.

An effective management of the disposal process will minimise surplus holdings and under-performing assets and maximise the Return On Investment (ROI) of the portfolio.

Functional Service Silos and Vertical Processes

IAM includes a functional review of all asset related services with special emphasis on Vertical Processes within functional areas. There are typically seven such functional areas: corporate real estate management, facility management, human resources, information technology, telecommunications, finance and strategic management, and customer operations. IAM provides a detailed analysis of each functional area, with the intent of identifying best practices in facility management and strategies that can be applied to improve performance and reduce cost.

Integrating Corporate Services

While a review of vertical processes is important, the integration of work between these functional areas is one of the key requirements for a successful IAM program:

- Horizontal Processes (between functional areas): the identification of opportunities for consolidation and/or integration of personnel, processes materials, etc. between functional areas.
- Integrating Functions (combining support functions).: the evaluation of management systems and other processes that apply to all functional areas.

Client Relationship Management

For the IAM Process to work, their must be a client relationship management program with the senior management of divisional or business unit leaders. This requires an understanding of each business and:

- An interface with senior management from all business units to develop major business strategies and real estate/asset initiatives.
- A consolidation of like business functions to reduce redundancies, foster synergies, promote re-engineering.
- The elimination of phantom planning and design work.
- The confirmation of occupancy-direct staff, full-time equivalent (FTE), consultants, and vendors.
- A focal point for organisational space engagement processes with senior management regarding significant changes planned by the businesses.

7   FM Performance and Accountability

John Hinks, Centre for Advanced Built Environment Research

FM in the Future

Any discussion of how facilities management might be evaluated in the future, risks making great presuppositions about the stability and continuance of FM as a recognisable and distinguishable profession. This uncertainty lies at the heart of the issue of what the foreseeable future of FM will or could look like.

Consider first the future use of the term 'facilities management' - will the term 'facilities management' continue to be ascribed to a multiplicity of definitions? Perhaps FM will converge into a singularly definable activity, or service? Perhaps the panoply of activities enveloped by today's headline term FM will consolidate into a number of distinguishable clusters of activities, each characterisable by common denominators such as the nature of service provision; the management or service process; the application of technology as the service (or in support of the service); the contractual protocols or the particular genre of business needs? And, if so, should that be Facility Management, Facilities management or facilities Management? Or all of them? All of these issues and questions will affect how FM performance is assessed in the future.

Secondly, consider the context of FM. Would an assumed ongoing stability in FM originate from a stable definition(s), stable ways of working, stable relationships or inter-relationships within the organisational structure, or from a stable scope and boundary to FM activities, or perhaps the stable structure of the industry itself? Furthermore, is this dynamic stability or static stability? Or, again, all of them? These first two concerns relate predominantly to the internal architecture of FM. The future of FM probably hinges more upon the external perspectives of FM held by the business world.

Thirdly, consider the continuance of FM. It originated from a number of sources and has many possible futures. Probably not as many possible futures as it had at one time, but still many possible futures. This is a pivotal assumption when future-gazing on performance assessment.

So, bearing in mind that the early evolutionary history of FM is characterised by turbulence and adaptation, is it reasonable to assume that the possible forms of FM in the future will be recognisable when viewed from today's perspective? And, notwithstanding this, will the criteria used for performance assessment be recognisable? Perhaps yes, and therefore probably yes again; if only because business will have to recognise FM in the future as emerging and growing from the FM of today, if its maturation is to be recognised and its future secured. This shifts the emphasis on future-gazing about

FM into the realm of business development trends, which in turn raises two clear dimensions of business need into the future. The case needs to be made that it is these dimensions which FM should aim to encourage for the future focus of performance assessment – a step which could also accelerate the maturation of FM from the clients' viewpoint.

The first dimension relates to the nature of recurrent and continuing needs for FM, which may change in terms of mode and form of delivery, but are likely to remain essential fundamentals to the operation of business. Those portions of FM which support the existing and continuing needs will have to refine their service. They will probably develop into the future by means of the clustering of activities and the attainment of best practices which itself will begin to change over time incrementally. A stabilising of FM in other words. Here the success criteria for FM may relate to continuous improvement, reliability, consistency, cost control, quality and time. But this scenario ultimately invites stagnation in the development of novel approaches, and was one of the drivers for the emergence of FM from previously stagnated and unrelated areas of activity. What goes around could come around (again) if the future view is not robust.

The second dimension relates to the developing business needs. Here success criteria for FM are likely to be associated with innovation. High cost may matter less than speed of delivery. A bespoke attunement of the FM service to the user's individual competitive needs, particularly in industries where distinctiveness and agility of process as an element of competitiveness could be more valuable. This places the emphasis on FM fundamentally in the arena of business innovation. Here the predominant features of good FM in the future may major on value, adaptability, novelty, support for new processes or timeliness. The balance of importance will of course vary between each organisation, so too will the criticality of support service supply, so also should the Key Performance Indicators used to assess the performance of assessment. Herein lies the performance assessment challenge for FM.

Consider then, what is it that business needs now, and will continue to need in some possibly transmuted form into the future, asking which of these needs are critical to the successful operation and agility of the business and are dependent on the appropriateness of its support service? Most facets probably, which does not help a great deal in narrowing down the key performance indicators or how they interact to confer efficacy on the FM service. Bear in mind that there are two dimensions to this issue also. The first, the more obvious, is the changed need of a changed business of the future. That places the emphasis on the future as a goal. The other, which FM must immediately address, is the changing need of a changing business world. Within all of this is the problem that not all businesses are equally adept at planning and managing their core through change. There appears to be the likelihood that a high performance, proactive and strategically dynamic FM service may exceed the vision and capability of some sectors of business to use it. And with this comes another type of complication for FM performance assessment.

Another dimension which may be helpful to distinguish the FM of the future is how it is used by business. Consider the support service context of FM. The business needs for support services require different priorities and hence different weightings on facets of FM for different business purposes. Contrast, for example, what would constitute usefulness or efficacy of the FM requirements of a business in the following modes:

- Restabilising immediately after changing its core process/production; or
- When in the act of changing its core process/production; or
- When considering the options, or planning the change process; or
- When it has a stable core process which it simply wishes to continue supporting; or
- When it wishes to protect the consistency of its core process/production from external pressures by the creative adjustment of its support service.

The performance drivers of FM would differ in these modes, perhaps some drivers would not even apply to all circumstances. This approach may be an appropriate way forward for distinguishing FM in the future – noticeably less self-focused and less preoccupied with what it is – and more focused on how it is used by the business, leading to the performance assessment of FM in the future being based upon how useful FM is to the business.

However, the ultimate usefulness of FM may derive from the contextual relevance of the levels and type of service provided, which that is the combination of a number of factors that really matters to business rather than the factors taken individually. That is not to say that individual factors found to be in deficit may not be critical to the success of the combine and the support for the core business, but rather that the value lies in synergy not simply singular contribution. This could complicate performance assessment in the future, since the current approach of reducing performance into individually measurable inputs based on today's vision of FM activities tends not to promote the view of a co-ordinated and strategically-sophisticated service. So the challenge for the future of FM performance assessment may be that it will have to involve the measurement of the whole rather than merely summing the parts. Bearing in mind that this correlates with one of the key dimensions of FM growth and added value, the challenge is really on 'measuring what is important', not continuing the mistake of placing the emphasis on the 'importance of measurement'.

The holistic view of performance also illuminates the individuality of the specific business circumstances that FM finds itself servicing. It is essential to consider performance assessment at the level of the specific service provided – since the added value of FM resides within the interactivity which it achieves for the individual business circumstance. The dynamic combination of the individual facets of FM performance, and the relative weightings between them may vary between the differing circumstances outlined above. Moreover, since this variation may occur for the same business over time as it experiences the differing episodes of FM need, there is unlikely to be any singular formula for assessment overall. There is little good news for benchmarking either – the variance between businesses will be qualitatively greater and more complex than within a single business, even assuming that the external environment has the same relevancy to each business.

So, taking all of these complexities into account, is there anything solid that can be said about the way forward for FM performance assessment? Is the challenge of defining and measuring what elements of FM, when combined, confer competitiveness on the core process simply too complex and arduous to be worthwhile? For most businesses, the answer is probably yes. It seems to have been so to date. But for FM the stakes are the chance to be part of the business change agenda for the future. For FM collectively it is certainly worthwhile.

Within the scenario developed above lies the paradox of measuring the value of an integrative FM service which is as complex yet invisible as the business interactivity, and is complicated further by its own operational and managerial processes. Is this just simply too complex and too much for the business to be concerned about? Is it too ethereal for all but the best and most far-sighted FM organisations to tackle? Can either party approach it in isolation, or does the future of the performance assessment of FM lie in joint strategic initiatives between FM and the core business strategist? If so how does FM, which has clearly evolved by filling the voids between the other elements of the business process demonstrate its value? Simply put, the nearer FM comes to achieving the strategic usefulness and operational integration that the business could benefit from, the more impracticable it becomes to distinguish it from the core, and hence to measure its contribution to the overall business outcomes. And all know that if it cannot be measured it cannot be managed! So is the problem of measuring what FM could mean to the business outcomes an unfathomable paradox?

Useful FM

Perhaps there is scope to assess FM performance using an alternative approach, whilst still preserving the insights into the integrative value of FM. Is a generalisable, benchmarked approach meaningful or does this actually divert attention away from the value of integration and contextual suitability of FM, which is where its real value lies? Are we in danger of measuring the wrong things if we continue to use the contemporary benchmarking data, and especially if we move towards generalised data sets instead of targeted models of FM usefulness?

Companies which continue to adapt and grow within their changing environment will probably only be able to do so by having the resources they need. The adaptations may be in support of the sorts of modes of changes presented earlier. Hence the particular needs, and so the particular balance of support, will be highly variable. Accordingly, the balance of potential usefulness to the business from reactive and proactive FM, also from strategic and operational FM, will differ. For instance, a business which is operating a stable process, and in an unchanging state may be able to make most use of an operationally efficient and cost-minimised form of FM. Facility usage and demands on other aspects of the business support services may be stable and align well with the operational and facility-oriented definitions of FM. The FM need may be predictable and stable. Little investment may be planned for the purposes of change, and a reactive rather than proactive FM with little strategic input to the business may suffice and represent value in these circumstances. For this type of organisational situation, the performance indicators may remain broadly the same as they are now. However, this sector of FM provision is not where the future growth of FM will major. Hence it is not the whole story for the future key performance indicators of FM.

Consider the same business in the options-analysis stage of planning a change in their core process or production/service provision. Here the adaptability of the processes may be more critical to the future competitiveness of the business than the actual cost of the FM needed to ensure that provision. Here the adaptability of the core processes may be more critical to the future competitiveness of the business than the actual cost of the FM needed to ensure that provision. Proactive FM foresight and a strategic understanding of the inter-relationship between the business and the FM may be very valuable to the business, and any stagnation or unresponsiveness caused directly by the nature of FM provision and outlook could severely limit its usefulness to the business during the change phase and the changed phase. Such unresponsiveness may limit the change options, or potential speed of change, and the efficiency of change and the changed business. It may dull the organisation's ability to respond to a turbulent environment. Note, however, that in addition to any problems with the inherent capability of the FM provider, such limitations may derive from the nature of the contractual arrangements for the provision of the service, and/or the remoteness of the FM service from the core business. Clearly then the measurement criteria and key performance indicators would differ for the same business and same FM organisation in these two scenarios. They would also require to be highly contextualised for a business to be able to strategically optimise the FM usage, let alone to be able to compare the outcomes of FM performance with others.

Consider as a third example the adaptability of FM per se. This may be crucially important for an organisation which is trying to protect the stability of its core process or production via changes in the nature of the facilities it uses for this. This is a situation facing many organisations in global markets as they downsize or rationalise their global facilities needs. The adaptability of the FM service may also be absolutely pivotal in the post-crisis adaptation of the core business process, the dependency on which could arise during restabilising after a move or other structural re-organisation of the core business, or could occur as a result of an unforeseeable disaster. Here cost efficiency is likely to defer to other factors such as operational effectiveness, the robustness of the transitionary or new service provision and alignment and the quality of FM service alignment with the re-stabilising needs of the core business. Through this, the future debate over the contribution of FM to business can enter the realms of 'good' costs (Olve et al, 1999). An FM organisation which does not understand the nature of these needs (or in the context of emergencies cannot predict and pre-plan for the needs) is unlikely to achieve the required level of usefulness to the business. Once again the definition of FM, and the key performance indicators will differ in terms of the weightings placed upon them. This may also suggest the need for FM to be divided according to what its priority role is in the context it is being viewed in. It also suggests that a singular FM provision designed for one mode of service need may contain too many critical compromises to allow it to achieve the required performance profile in another mode. Evidently the value of the outcomes depends somehow on the balancing act required in the process.

Key Elements of Useful FM

It would seem to be impracticable, or of little real value, to try to produce a strict set of parameters for FM provision which are aimed at the generalisable supply issues within FM rather than the demand-oriented issues. Besides which, as Rumelt (1994) observed, products and services are only a temporary manifestation of core competencies, and these data don't support the management control of FM, rather they focus attention away from it and onto the facilities. Nevertheless, this is currently what is being done by some for assessing their FM performance. However, are there any common or recurrent demand factors which could be identified as key performance indicators that could comprehensively represent the potential range of business needs in the future? Clearly, the idiosyncratic needs would have to be respected in their application, something which would surely need the joint consideration by the business customer and the FM provider. Within this scenario, it looks likely that the individual provisions that are needed in different circumstances are unlikely to relate directly to the inputs to FM, and in many cases will not ultimately be measured in relation to the FM process, although there will be intangible, and possibly tangible, interactivity and inter-operability issues at the juncture of the core business and FM. Hence the value and also the measurement of performance may be more appropriately considered in terms of the robustness or efficacy of the FM supply to meet the business demand. The usefulness of FM in this context is also likely to be assessed in terms of outcomes (not necessarily outputs). These outcomes will be the result of complex and systemic inter-relationships between FM and the changing business process. Strong co-operation between the FM and the business will confer agility and adaptability on the business whilst preserving resilience and robustness of the core process.

Through this perspective, FM may come to be seen as a performance driver. But usefulness may ultimately depend upon a combination of support service provisions. This raises questions over the concept of core and non-core distinctions during transition. Clearly, whilst some aspects of FM will remain non-core in some circumstances, some features of FM will be less consistently non-core. Hence the key performance indicators for FM have to be bespoke and supply oriented, how is this usefulness to be appraised? Usefulness hinges upon the context of business demand.

Table 2 shows a list of key performance indicators for FM (Hinks and McNay, 1999). The list was systematically generated by a balanced group of FM providers and business customers and is presented in their order of priority. It includes a mixture of FM inputs to the business operation and to business outcomes, and it should be noted that some indicators do not help to directly assess the outcome impact of FM on business performance.

As a test of the universality of 'FM Usefulness', consider the factors in

Table 2 and decide whether it is possible or not to define how a level of performance equates to the industry norm (whatever that may happen to be). Next consider how these factors would confer usefulness on FM for a business, and then rank them for the purposes of performance management.

Table 2   Key Performance Indicators for FM.

- No loss of business due to failure in premises services
- Customer satisfaction
- Completion of project to customer satisfaction
- Provision of safe environment
- Effective utilisation of space
- Effectiveness of communication
- Reliability
- Professional approach of premises staff
- Competence of staff
- Management of maintenance
- Responsiveness of FM dept. to changes and requirements
- Value of money
- Satisfactory physical working conditions
- Equipment provided meets business needs
- Suitability of premises and function environment
- Quality of end product
- Effectiveness of helpdesk service
- Achievement of completion deadlines
- Correction of faults
- Standards of cleaning
- Management information
- Energy performance

Well, it is probably almost completely impracticable to do so without more information about the context of the need, excepting the indicators which relate to business continuity and customer satisfaction. There may even be circumstances where these issues are not clearly a priority.

So the value, or usefulness of the other factors cannot be appraised without some more contextual detail, and this is the central point of the concept of assessing the performance of FM in the future. It is a complex matter of balancing multiple objectives (Olve et al, 1999). If FM is to prosper it has to be measurable, and hence defined and manageable. But before this can happen in a comparative sense, FM is going to having to consolidate into various forms and contexts, so that the relative criticality and weighting of the services within FM can be measured and compared appropriately for different contexts. More importantly, but probably contingent upon the former refocusing on the value of what FM contributes to business outcomes, are the bespoke demands of the business in its current mode of dependency on facilities. In these circumstances the essence lies in usefulness, which is thoroughly context-dependent. It is therefore probably reasonable to expect that the evolution of performance assessment of FM in the foreseeable future will become inextricably bound up with the modus operandi of FM in the future. Hence the future of FM and the future of performance assessment of FM are entwined.

The FM Debate

This chapter has speculated on the future of Facilities Management in the context of the future of business needs. It has observed that the focus of contemporary FM performance assessment on the verifiable has limited the consideration of the wider, perhaps less tangible or differentiable value of FM. There is currently a mismatch between the performance indicators preferred by the FM industry - which tend to be reductive databases designed for the quantitative comparison of the facilities-oriented aspect of FM - and those performance indicators which the core business is interested in, which tend to be more synergistic and business-outcomes oriented issues. The emphasis on measuring FM performance separate from the business has also neglected its interactive value. This is a difficult issue to tackle, but one that lies at the heart of the contribution that should be assessed.

This chapter also suggested that a shift in emphasis will have to occur in the foreseeable future towards the assessment of the usefulness of FM, using business-oriented criteria for the quality of the business support service. As Rumelt observed (1994), "in the long run, competence, not products, will determine who succeeds in competition". Perhaps within this usefulness perspective, the dependency of the core organisation on the FM outcomes and the opportunity costs of under-performance may be worth considering to give a multi-dimensional view? Considering FM from the business view, the development of models which would be suitable for assessing the business usefulness of FM may need to rely upon techniques such as the Balanced Scorecard Approach (Olve et al., 1994), or the recent FM-specific Management by Variance approach (Hinks and McNay, 1999). By taking a developmental approach based on the business interest in FM, this may help achieve the level of understanding needed to get wider business uptake of FM. Meanwhile, there is currently a mis-match between the performance indicators preferred by the FM industry – which are tending towards generalisable, reductive databases for benchmarking the facilities-oriented aspect of FM – and those performance indicators which business is interested in, but which tend to be more synergistic and business-outcomes oriented. The reductive key performance indicators continue to lie at the heart of assessing the provision of facilities-oriented services. However, if the current trend towards facilities-oriented, generalisable, and reductive data continues unattended, rather than supporting the evolution of FM into a strategic lever for business advantage, it would serve to normalise and hence obscure the potential of FM to contribute to business competitiveness on a bespoke basis.

It appears likely that the future competitiveness of a significant proportion of the business world will be pivotally dependent upon dynamic rather than static competitiveness (efficiency of process and production of a stable product). In such circumstances, relative business advantage will be derived from the strategic application of FM in a customised manner, wherein the business support service is designed and assessed according to the specific strategic needs of a changing business. The indicators for assessing this aspect of FM performance will have to be more high-level, more transparent to business and will have to represent the synergistic outcomes of a range of inter-related FM services, probably the result of a complex interactivity within and with the core business. Accordingly, the chapter speculated that, in the increasingly globalised business market, the combined variability of political, economic, social and technological environments and internal-to-business responses to these environments will mean that the usefulness of FM to a large portion of the business world will ride upon the scope of FM to adapt and achieve synergies in conjunction with the core business processes (designed and appraised in relation to the idiosyncratic needs of each individual business).

In these circumstances, the target business value of FM may reside in its usefulness when applied strategically for competitive business advantage. Hence the key performance indicators for FM of the future may have to be directed more towards business outcomes, and could converge with core business performance indicators such as agility, flexibility, business continuity, and/or transition management. For changing business, the nature of FM needs and priorities within these needs will also change, and hence the balance between facets of FM support will change as well as the levels required of a particular service. Furthermore, where there are tensions over the balancing of resources for service provision, a flexible approach to interpreting the criticality and contribution of individual aspects of FM will have to be co-ordinated with other decisions based on performance measurement. For businesses that currently do not appreciate the potential usefulness of FM, this would also represent the opportunity to establish current strengths and weaknesses of business accruing from the contemporary nature of support provision and integration.

This raises two important issues for the contemporary emphasis on a reductive evaluation approach to FM. First, that the input of FM will no longer be seen as directly relevant to business assessment of high-level FM performance. Secondly, the pressure will be on outcomes (representing a further development of the current trends towards performance-oriented Service Level Agreements). These outcome premeditated factors may need to be balanced with input factors via trade-off decisions which recognise the interdependent nature of FM provision. The same may apply where performance assessments are used in relation to ROI assessments for best value.

Overall, these factors may stimulate a polarisation of the FM industry into Facilities management, characterised by a predominantly operational service supply industry (probably also predominantly provided through outsourced specialists offering similar types of service and competing on cost or value) and facilities Management, characterised by an in-house strategic, consultancy-oriented, mediating form of management. The suites of key performance indicators required for publicly assessing each form of FM in the future will be significantly different, and through the maturation of performance assessment to allow the strategic comparison of FM impact on the organisation to be understood, FM will emerge the stronger and the more useful. Then the separate challenge of using this for the management control and maturation of FM will really start (Hinks, 1998). As businesses start to recognise the relative competitive advantage to be derived from an early uptake of strategic FM, so this perspective on performance will assist FM to gain a corporate foothold. The question still remains over whether this will still be known as Facilities Management in the future.

References

Hinks J. (1998) A Conceptual Model for The Inter-Relationship Between Information Technology and Facilities Management Process Capability, *Facilities,* Vol. 16, No. 9/10, Sept/Oct.

Hinks J. and McNay P. (1999) The Creation of a Management by Variance Tool for Facilities Management Performance Assessment, *Facilities* Vol. 17 No. 12, Jan/Feb, pp. 31-35.

Olve N.G., Roy J. and Wetter M. (1999) *Performance Drivers – A Practical Guide to Using the Balanced Scorecard.* Wiley, Chichester.

Rumelt R. (1994) Hamel G. and Heene A. (Eds) *Competence Based Competition.* Wiley, Chichester.

Part Two

The People Trail

8 Introduction to the People Trail

For more than thirty years fundamental changes to the nature and organisation of work and have been forecast and their likely impact on the demands for building space predicted (Cowan 1969). Over the last decade these changes have begun to happen visibly. Today, change at an unprecedented pace, is leading to quite different organisational structures and priorities, new and more varied employment arrangements with the rapid diversification of work practices and venues. As a result, some of the long established approaches to human resource (HR) management and its workplace support are becoming inconsistent with the needs of modern business. The consensus of the late 1990s has been that these changes are driven by:-

- Information Technology: the impact of telecommunications, networked PC systems, the internet and cordless technology, expanding the ways in which businesses communicate one with another, and fundamentally changing locational constraints and opportunities for organisations.
- Organisational Requirements: distributed business operations to subcontractor and staff, normally requiring smaller and more flexible units of usable space and shorter leases.
- Work Practices: flexible working becoming a more integral part of HR strategy in many organisations, regardless of sector, with increased reliance on space sharing, multi venue working, home working, teleworking, 24 hour working, etc.
- Employment Strategies: more varied, adaptable and shorter contracts of employment, including flexible working hours, 'zero hours' outsourcing.
- Employee Expectations: increased concern for amenity, health, safety and well-being in the workplace, particularly the demand for more natural and controllable work environments.

But what is the true position regarding these changes in work patterns and practices? There seem to be two contrary views. The first is that the importance of changing working practices has been overstated, their impact greatly exaggerated. Those with this view suggest that change will continue to be gradual and evolutionary - the cautious position - in which the future will be much the same as the past, at least in the short to medium term. The second view is that irreversible changes are underway that will profoundly alter the nature of work and the support facilities that are required. This view suggests revolutionary change; the future will be quite unlike the past.

The debate about the possible benefits and risks of these changes is ongoing. At this early stage in the introduction of new ways of working, the possible advantages and disadvantages, are not fully understood nor has the claimed contribution of flexible working to business performance been thoroughly tested. Some hold a positive view of the potential benefits of flexible working (McLennan and Cassels, 1998), others adopt a more conservative and questioning approach (Cairns and Beech, 1999). However both the adventurous and the cautious recognise that changes of this kind are giving rise to quite different facility requirements, particularly in the office, retail and residential sectors. The changes, and the uncertainties that they bring, are placing more exacting demands on the planners, designers and managers of the workplace and are raising questions about the logic and validity of conventional briefing, space planning and allocation procedures. Many established benchmarks, space standards, utilisation targets and 'good practice' procedures are becoming at best inappropriate, at worst redundant. More critically, these changes demand new professional skills and approaches, both for those that manage human resources in the work environment (Pettinger, 1998) and those that are responsible for the management of facilities and support services.

Figure 18 Work Organisation in Space and Time.

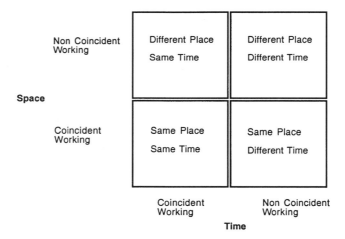

How can these fundamental developments along the People Trail be properly considered? There is a simple but basic starting position for such discussions. This relates to the ways in which work and work communications can be organised in space and time (Nutt, 1992). Work can be conducted in one of the four basic circumstances shown in Figure 18, or some combination of them. First, work with others may be organised to coincide both in space and in time. This is the 'face to face' workplace of the traditional office and is still by far the most common way in which business is conducted. Second, work can be arranged to be coincident in space but not necessarily in time. 'Touchdown' or 'drop-in' facilities that support multi-venue working arrangements are an example here. Third, work with others may be coincident in time but not in space, for example, satellite office working with video conferencing support. Finally, and only recently possible, work can be organised to be undertaken in circumstances that are neither coincident in space or coincident in time; this mode of work underpins the 'virtual office' and many global working practices. These then are the four basic options for organising work in space and time:-

- Same place - same time;
- Same place - different time;
- Different place - same time;
- Different place - different time.

What is unique about this contemporary work environment, is the increased diversity and variety of working practices, work locations, venues and types of workplace, across the four domains of work shown in Figure 18. There are three key dilemmas here. The first concerns an organisation's choice of an appropriate 'mix' of the four working domains to suit the requirements of its business and the needs of its employees and their work. The second dilemma concerns the choice of management structure, procedures and culture. While there is extensive HR management, line management and facilities management 'best practice' experience within the traditional 'same place - same time' box, expertise and methods of approach are at an early and mostly untested stage of development within the other three domains of work. The third dilemma relates to the quantification of the demand for, and the supply of, facilities, space and support services. Here again, estimates within the 'same place - same time' box can be securely based on proven methods. Estimates in the remaining three work domains are problematic with few available approaches and techniques and no management protocols on which to rely.

Figure 19   Space-Time Profile

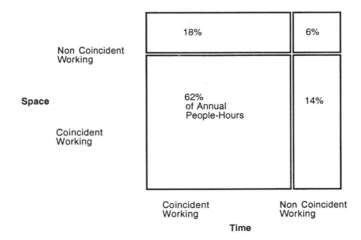

As introduced in Chapter 1, the primary resource to be managed across the four domains of work, is the total people-hours of employee time per year that is available to an organisation. Figure 19 shows a hypothetical situation in which it is estimated that approximately 62% of the annual total human resource (measured in people-hours per annum) will be allocated to traditional 'face to face' situations, 18% and 14% to 'same time - different place' and 'same place - different time' activities respectively, with only 6% of the total people hours allocated to 'virtual office' type working. Naturally such allocations will change over time. Predictions of the allocation of any individual's time budget, or for the allocation of the HR to be deployed on any given project, between the four working domains can be illustrated in an identical way. Flexible work profiles of the kind, as shown in Figure 19, can be used to indicate the quantitative demand for the variety of facilities and services to support the HR of the business as a whole, its sections, its project groups and for individual employees. This is only a simple beginning. It is the more qualitative and complex issues of managing organisations, their employees and facilities within and across each of the four domains of work, that is most problematic. Key questions here include:-

- How will the changing patterns of work be supported in the future, particularly those that are not undertaken at 'same place - same time' facilities?
- What are the likely consequences of flexible working arrangements for HR management practices?
- What responsive management approaches and techniques should be developed for infrastructure management and logistics in support of an organisation's human resource deployment when flexible working arrangements are adopted?
- What types and scope of facilities management support systems will be required so that the promised benefits of flexible working may be realised and their potential risks contained?
- How might the effectiveness of remote and distributed facilities management and service support delivery be monitored ?
- How will individuals and project teams adapt to operate within a flexible working environment and what new forms of stress might flexible working entail?
- What are the likely social and cultural implications for the individual and the organisation?
- What new types of facility and innovative work settings might be designed and developed to meet the changing requirements of work?

It is questions of this kind that are addressed in the next five chapters of this book, with each chapter exploring possible FM developments from a predominantly human resource perspective. The next chapter on the People Trail provides a comprehensive overview of a future where work will no longer be a place, but a range of activities that can be conducted virtually anywhere and at any time. It details the critical management issues that need to be considered when providing a responsive and effective human support service to underpin a wide variety of transient work patterns, undertaken at different venues and at dispersed locations. Chapter 10 develops these ideas further and considers their implications for workplace variety on the one hand, and for employee empowerment on the other. The potential conflicts between the objectives of an organisation and the needs and wellbeing of the employee are considered. A future where complex combinations of traditional and flexible ways of working predominate is predicted. This future will demand a higher level of facilities management skill than in the past, with measures to evaluate the effectiveness of workplace support to individuals and teams, superseding simple measures of efficiency alone. Here, not only the technological, legal and facility implications of flexible working need to be considered, but also the significant social, psychological and cultural changes that seem to be underway.

Chapters 11 and 12 examine the increasing diversification and choice of workplace support systems and locations. Chapter 11 makes the case for a future where self-selection of working mode and the self-selection of appropriate workplaces, by teams and individuals, becomes commonplace. It is argued that developments of this kind will also involve the delegation of many FM functions to teams and individuals, radically changing the ways in which facilities and support services are managed. Chapter 12 explores some of the key psychological and social dimensions of flexible working and the physical workplace, both positive and negative. A 'stakeholder' position is proposed with a new balance being struck between the concerns of the organisation, its business and customers, and the needs of individual employee, working teams and their workplace environments. A future based on a participative 'democratic' approach to workplace design and management is advocated, with facilities management, as its name implies, expertly 'facilitating' the participation process.

Finally, the concluding chapter on this People Trail looks at the management of the journey to and between workplaces, and the measures that can be adopted to reduce stress and non-productive time. It provides a case study account of the development of a Commuter Plan and corporate commuter management arrangements to facilitating the transport of an organisation's human resource to and between work venues. It summarises the experience gained and points to a future where the scope of FM in supporting people may move beyond any specific workplace location to include wider support arrangement concerning travel, recreation, employee life-style, family and community concerns, and other issues that can be shown to contribute to staff well-being and productivity.

References

Cairns G. and Beech N. (1999) Flexible Working : organisational liberation or individual straight-jacket, *Facilities*, Vol. 17, No. 1/2, pp 18-23.
Cowan P. et al., (1969) *The Office: a facet of urban growth*, Chapter 10, Heinemann Books Ltd.

McLennan P. and Cassels S. (1998) New Ways of Working : freedom is not compulsory, *Facilities Management - The New Agenda, BIFM Conference Proceedings*, BIFM Conference, Cambridge, pp. 72-76.

Nutt B. (1992) Chairman's Opening Address, *Flexible Working: The Changing Environment*, School of Business and Industrial Management Conference, London, 18 April, material based on MSc FEM course notes, event 5.6, UCL, London.

Pettinger R.,(1998) *Managing the Flexible Workforce*, Cassell, London.

9 The Future of Workspace Management

Wes McGregor, Advanced Workplace Associates

The provision of workspace is, or at least should be, a direct response to the considered needs of people (individually and collectively) in supporting them in their work endeavours (current and future). Consequently, to ensure that the most fitting work environments are provided to meet their needs, those tasked with the responsibility of developing workspace strategies and planning workplace layouts must be constantly considering their needs for the years ahead.

We live in an information age, where work is being redefined beyond traditional bounds of space and time. It follows therefore that as the requirements of work itself changes, then so too will the requirements of the management of workspace. Inevitably its future will be influenced, if not determined by what the future holds for the component parts of the space management process, namely – work and work processes; the people who comprise the workforce, their lifestyles and expectations; the technology of work, affecting both the users of the workspace as well as the processes of its management; and the property industry and their responses to the dynamic needs of those they serve.

The old way of planning workspace was to oversize the amount of space in the knowledge that eventually as the volume of work increased and head count grew with it, then the initial excess of workspace would be absorbed. This is not an option in the 'trans-millennium' age we currently inhabit. What if the organisation needs less space, not more? What happens if the space that is required is needed in a different location? Or even, that no space is required at all? The principal difference is that today, and increasingly in the future, the nature, extent and speed of change is greater than ever before. In an age of rapidly changing business needs, new ways of 'employing' people and a different geography of the marketplace, facilities managers can no longer rely on the inevitable increase in headcount to absorb their planning mistakes. Add to this the pressures most organisations face for reducing costs especially the unit cost of production in both manufacturing and service industries, and the need to reduce the reliance on, and impact of, fixed costs such as rent and rates, then it is clear that facilities managers and their businesses need to consider their workspace in ways different to those of the past.

In 25 years time the world of work will look very different from the way it looks today. The future visioning of the likes of Stanley (1995), Handy (1995 and 1996), and Boyett (1996) and Bayliss (1998), suggests that the future world of work is likely to be characterised by features such as those listed below, the evidence of which can already be seen today:

- Fierce international competition will further squeeze product design to delivery cycles, resulting in demand for facilities that must go beyond the fundamentals of quality, reliability, cost, and even beyond flexibility, to agility – the ability to make anything, in any volume, anywhere, at any time, by anybody.
- Work environments that are barrier-free settings in which people are able to work intuitively, where they can organise information and interactions, and where they can choose the location, pattern and timing of work.
- In response to their requirement to reduce costs and to use resources be they property, technology and especially people, just when and where they are needed, organisations will embrace a variety of people – core employees, self-employed, partners, freelance contract labour, subcontractors, and e-workers – in an inexorable move towards virtual organisations who embody the whole supply chain.
- The availability of powerful information technology and communication services to control and manage effectively distributed and remote business operations, irrespective of geography or time, enabling workers to be globally mobile without leaving home.
- Higher quality standards and lower cost will not be seen as mutually exclusive but as vital and mutually complimentary necessities of successful businesses seeking competitive advantage.

Many workplaces of the 1990s have been designed as responses to old approaches to work, and in so doing, fail to take account of the present needs of people and businesses, let alone their future needs. This will have to change as organisations will be far less able, let alone willing, to tolerate work environments that are not fully supportive of their work endeavours. For many facilities managers such a world may seem far removed from that which they inhabit and serve today. But what is clear is that work is going to change, almost certainly irrevocably, in terms of the how, when and where. It is against this backdrop that we look forward 25 years and consider the role of workspace management in terms of these key features and the issues they raise for those charged with the responsibility of managing their organisation's workspace.

Workspace Strategy All successful, i.e. healthy and sustainable, enterprises will have 'living' workspace strategies that evolve as the organisation evolves. Like their more conventional counterparts these strategies will embody the what of business in terms of the activities, but unlike them they will now wholly embrace the how of business i.e. people, their interactions, tools and processes. The place of work, i.e. the where, will at one and the same time become less relevant and more important. Less relevant because the capabilities of technology will blur the traditional distinctions between work environments and buildings permitting work to be done anywhere, since the cyber environment will be equally as important as the physical environment; more important since those environments that cannot sustain this 'free-style' working will be rendered obsolete and condemned to the demolition ball. In other words, strategies based upon the economic realities of the occupying businesses and not merely of the traditional short-term horizons of the property industry.

Flexibility will be the hallmark of effective workspace strategies where most will encompass a combination of core and elastic workspace, in their attempts to address business volatility and fluctuating levels of demand. The elasticity being obtained from a variety of propositions that include serviced workspace available for various and varying durations to suit the fluctuating demands of the business, with some being provided by a mix of 'corporate centred' information stores, meeting spaces and home environments. Serviced workspace previously seen as the domain of the small start-up enterprise who could not afford a 'place of their own', will be at the heart of alternative accommodation strategies for many businesses, as they adopt a virtual approach to workspace. The emphasis will be on the sharing rather than the owning of space by individuals, work groups, functional departments, and even by businesses and government departments – leading to an increase in demand for high-quality, flexible workspace, available on a variety of flexible commercial terms. The challenge for facilities managers is how both new (and more often) existing building stock can be transformed, used, and managed over time to meet these markedly different needs.

Workspace Location

As risk is diversified away from the main downstream producer, rather than working from one or a few locations, businesses will typically operate from a great many locations, progressively evolving towards virtual and varying organisations – using premises of customers, suppliers, partners and JV alliances. In this form the business becomes a network of relationships with other organisations in which it can readily and speedily access specialised know-how on best practice operations, wherever they may be located, in order to create high-quality, low-cost, new products and services for customers on demand. Consequently employers will focus on the attainment of a specified output rather than their traditional role of someone who demands attendance at a fixed time and place. Most workers will split work time between a range of locations which are likely to encompass a central hub facility, dispersed satellite premises, telecentres, business centres, hotels, transport nodes such as airports and rail stations, transportation systems themselves such as aeroplanes and trains and of course home environments. The successful management of these disparate environments will call for a diversity and depth of skills, knowledge and resources not currently possessed by many facilities management teams.

The extent to which the rapid growth in electronic and satellite traffic, email and video conferencing, will change the economics of location is uncertain, but the motivation is likely to be quite different from current drivers. The use of these technologies permitting remote access to expertise (RATE) enabling businesses to provide specialist skills from dispersed locations rather than from concentrated 'centres of excellence', will question the need for centralised portfolios.

There has been for sometime a move towards decentralisation of 'backroom' functions to dispersed locations by some major business sectors like banking and financial services. Such moves are not only beneficial in terms of savings in facilities operating costs, but can also have the effect of improving the lifestyle of their employees through less time spent travelling to and from work. It follows that such initiatives will have profound implications for the locus of work, and resounding consequences for urban property markets, as the workplace will be location-independent and user-driven, fully functional and supportive, and occupied on an intermittent pattern by various people and for varying durations.

Environmental Response

Simultaneously, the effects of environmental pressures, the pervasiveness of IT and a re-evaluation of the quality of life, will combine to encourage many workers to explore new ways of working, as a means of liberating them from the personal effects of extensive commuting, stress, unproductive time, and soaring costs. Telecommuters have choice as they can chose to work before commuter trains start running or long after conventional workspace is closed. Many work activities could be conducted in dispersed satellite locations near home where access can be achieved with less stress, quicker and with less expenditure of energy. The triple whammy goes quadruple when in the home environment the worker is more productive than he would have been in the central workplace.

Environmental responsibility, driven by both legislative and voluntary initiatives, will continue to exert pressure for improved performance of workspace and buildings as eco-systems. All facets of the design and construction of work buildings will be selected after a thorough assessment of their environmental impact and performance targets set for their lifetime operation. Facilities managers will therefore need to have available to them reliable data about the operating performance of the buildings in their charge, robust indices against which this performance data can be compared and effective means of exercising control over all aspects of building operations.

Work Patterns

Erratic and often unpredictable patterns of working will be the norm rather than the exception, as businesses focus upon responding to the individual needs of customers. World-wide working on-line will be common hence the pattern of the working day, week, month and year will vary for each individual and from day to day and week to week. Premises and infrastructure will need to be available 24 hours a day seven days a week, to suit global business across different time zones as well as the personal lifestyle choices of people as they balance their needs for work and home. Such requirements can only be met through the provision of robust, resilient and controllable building systems, which will necessitate changes in the way investments are made in business workspace.

As the intensity of use of workspace increases, it will call for highly responsive maintenance and servicing, so as to take advantage of every small window of opportunity available, in order not to disrupt business operations while still maintaining business systems at optimum levels of performance.

Work Environments

A key objective of workspace management is to match the requirements of the tasks to be performed to the most effective time-space relationship that can be achieved. This will require a range of formal and informal work settings to be provided to support collaborative, solitary, individual and group activities. The business will require a 'central hub' or possibly several dispersed 'hubs', where workers can meet, share experiences, collaborate and socialise. The chosen setting may be a desk within business premises; but equally it may be on a train or aeroplane, at home or in a hotel, in a café or a conference room, in fact anywhere that is conducive to the type and timing of work to be undertaken. It is easy to see how such a scenario could be construed as the demise of workspace – shop, office or factory. This is unlikely be the case. While the overall quantity of conventional workspace is likely to reduce, new styles of working will necessitate an increase in the diversity and quality of work environments and their widespread location.

Within 25 years the growth of big call centres, so much a feature of the 1990s, will be all but gone as new channels to market via home based interactive technology are developed to replace them. This will lead to call centre agents working from home, or from places close to home rather than in aggregated central work buildings.

In the office market, with the growth of serviced workspace populated by transient guests who visit to perform specific functions, offices are fast becoming hotels for work. The users of these buildings will expect a more comprehensive range of facilities and services than have traditionally been provided in such places. Like hotel guests, workers require facilities and services fully at their disposal while they are in 'residence' before moving on to another work location. All of which is likely to result in an overall reduction in the amount of workspace that will be required to meet the collective needs of businesses.

In core workspace the process of making changes to the layout of settings has to be made easier, faster and less costly than at present, and under the control of the individual i.e. self managed change, through the application of developments like wireless networks. Such innovations will almost render churn to be a thing of the past as the barriers to moving people and workplaces is all but eliminated by location independent working.

Rampant advances in technology systems previously thought of being purely the domain of business will be available and accepted in the domestic environment. Houses will need to be built to meet the requirements of the home worker, through the provision of workspace, LAN, ISDN lines and other facilities. The home worker will be seen increasingly as a well-defined and separately identifiable segment of the housing market.

Supply-side Imperative

Buildings that are designed for single occupancy throughout their lives will be things of the past. In order to survive and flourish the property industry will need to cater for the real needs of varying occupiers, their varying levels and pattern of occupancy. In turn investors must recognise that the future performance of their property portfolios depend on their acceptance by users who will demand minimum operating costs and maximum flexibility as mutually inclusive ingredients for business success.

Office buildings typically take 2-3 years to develop; they are leased for periods of up to 25 years, sometimes longer, in which time businesses will have disappeared or radically changed shape. In the same time-scale new technologies will be developed, new products manufactured and new skills learnt. Some buildings can be adapted to cope, many cannot, or at least not without compromising the effectiveness of the occupying business.

Supply side imperatives therefore need to encompass:

- Funding and investment – as businesses are metamorphosing over shorter time-scales, then lease periods of 25 years and longer, are no longer relevant, let alone acceptable. Consequently other varied, and variable, types of funding and investment arrangements based upon 'partnering' between developer-owner-occupier, need to be provided.
- Whole environment score – which rates the whole-life of buildings on the totality of their impact upon the environment, covering such aspects as the energy and water they consume (in development and in operation), the levels of contaminants they emit (in development and in operation), their adaptability for other uses, the extent to which they make use of recycled products and materials and the extent to which the materials used can themselves be recycled.
- Total facilities management (TFM) – where the building, and all of its infrastructure, furnishings and fittings, are available for 'hire' via a composite lease or other form of commercial arrangement, all bundled as a 'managed' service.
- Life-time guarantee – where TFM is extended to cover the upgrading and/or replacement of all facilities throughout the occupiers' period of residence in the building, to meet their long-term variable needs.

**Workplace
Performance**

Competitive business pressures, continuing an unrelenting drive for the lowest possible cost of production, will demand ever-higher levels of operating performance from the individual and the team. Higher standards of functionality will be vital attributes of work environments where the workplace and its support services are operating under the direct control of workers and not of facilities managers. In such a situation the role of the facilities manager will be that of the 'enabler' marshalling the services required, keeping them under constant surveillance and continually seeking improved performance through the application of innovative techniques. This will in turn require the facilities manager to conduct frequent surveys of his customers (workspace users) to ascertain the level of their satisfaction with the work environment, its infrastructure and services. As businesses and facilities managers alike develop a better understanding of the correlation between the performance of the work environment (in its most holistic sense) and the performance of its users, workers will be much less tolerant than in the past, of defective and deficient workspace.

The management of workspace will therefore be as much about measurement as it will be about implementation. The facilities manager will need to possess a deep understanding of the way the workplace affects the performance of people (individually and collectively), how the workplace is being used and why and possess the ability to orchestrate changes in the way work is conducted, confident in his knowledge of the cause and affect linkages. By evaluating the performance of the workspace and linking the findings to recognised business metrics, then the facilities manager will be able to make a value added contribution to the business.

Facilities Management Operations

As the geography of the workplace changes, smaller cores of workspace will be required at the 'centre' of businesses giving rise to many and diverse types of workspace being used in a wide range of building types in widespread locations. For organisations with highly fragmented operations the critical mass required to sustain a dedicated and local facilities management team is unlikely to exist, so new initiatives will need to be developed in response to the changing scene. Joint ventures and strategic alliances will not only reshape the conventional supply chain models in favour of virtual organisations but have the potential to 'virtualise' the management of property and facilities. This will encompass methods and systems for remote operation, monitoring and diagnosis of facilities services. Through collaboration, joint venture and sharing the facilities management functions of neighbouring businesses will be able to combine and rationalise in support of a range of local buildings and their occupants.

The provision of facilities services, the customisation of workplace settings and the adaptation of workspace layouts, all need to be achieved much more readily than in the past. The facilities manager and his team, need to become custodians and providers of the means of production, the control of which is devolved to their customers. This is likely to require new forms of service delivery provided via various and variable forms of contract and more expansive help-desk services embracing remote diagnostic, the provision of the means by which individual control can be achieved, all featuring as integral elements of the selection of work setting components. All of which will have profound implications for compliance with workplace health and safety and other increasingly stringent legislative obligations.

To be effective and remain so, the work environment needs to be created from the collaborative efforts of all aspects of business, where IT and HR along with facilities management are combined to achieve an effective integrated support service for business operations. This will call for much higher levels of competence and business skill than traditionally have been evident in many facilities management practitioners. They will need to become 'multi-lingual' in terms of their ability to interpret the needs of their customers (individual, group and corporate) and translate facilities related issues into a language that enables them to communicate effectively with managers of the business. This should be seen as one step on the journey towards positioning the management of facilities as a business critical service.

Conclusions

Work is no longer a place – it is an activity that can be conducted anywhere. Consequently it is clear that businesses' requirements for workspace will change in terms of quantity, quality, location, diversity and functionality, all of which will place greater and more varied demands on those responsible for its provision and its management. To maintain an effective role in the organisation, facilities managers will need to innovate new solutions to new problems based ever more on looking forward and much less on looking back.

In response to the highly 'fluid' demands of business, the property industry needs to develop a wide range of products for the market which are likely to require new approaches to the way property is funded, designed, constructed, equipped and provided. The facilities services industry will need to shape new offerings based on more appropriate levels of service for businesses and their people, which are developed in recognition of the need to be value adding rather than resource consuming.

Accordingly managers of facilities must make a major shift of mindset by recognising that simply adopting coping strategies is insufficient, as the focus of their actions shifts from resource and expenditure containment, to adding value to the business. As they do so the role of facilities management will move closer to the centre of gravity of the organisation, helping to shape the business and its operations and in so doing contributing to its success. If facilities managers are to make an impact in adding value to the business delivery process, then they must aspire to a position of influence via their ability to offer facilities solutions to address business challenges. A prerequisite for which is a true understanding of the links between the quality of workspace provision and the performance of people and hence of the enterprise as a whole.

References

Bayliss V. (1998) *Redefining Work*. RSA, London.

Boyett J with Boyett J (1996) *Beyond Workplace 2000: Essential Strategies for the New American Corporation*. Plume/Penguin, New York.

Duffy F. (1995) The future for offices. *Flexible Working*. Vol.1. No.3. pp. 21-22.

Handy C. (1995) *The Age of Unreason*. Arrow Business Books, London.

Handy C. (1996) *Beyond Certainty*. Arrow Business Books, London.

McGregor W.R. (1994) An Integrated Workplace Process. *Facilities*, Vol. 12 No. 5, pp. 20-25.

McGregor W. and Then, D. (1999) *Facilities Management and the Business of Space*. Arnold, London.

Stanley J. L. (1995) Beyond flexible - tomorrow's agile manufacturing facility. *Facility Management Journal*. IFMA. May/June. pp. 16-20.

10 Effective Workplace Support

Sue Wood and Derek Worthing,
University of the West of England

To predict the future is notoriously difficult, mainly because just about the only thing that can be predicted is that the future will be more than a continuation of current trends. For example, extrapolations made not so long ago, for instance, saw the year 2000 as an age of leisure with a 30 hour working week (Thompson and Warhurst, 1998)! This chapter attempts to envisage possible futures by reflecting upon the reasons for recent changes in the workplace in the context of the inter-connections between the physical setting, the work and the workforce (one of the central tenets of facilities management), and the wider social, political and physical environment.

Amongst the many organisational and workplace changes that have occurred recently, much emphasis has been given to new technology and more flexible working patterns, often associated with the appearance of the term 'knowledge worker'. Many of the examples cited are, as Cairns and Beech (1999) have noted, in a fairly narrow organisational sector that has a vested interest in a 'showcase' (computer companies, marketing, advertising and design organisations). The reflections that follow make reference to the discussion of these issues in the national, business and built environment press and on the world wide web, but are also informed by more rigorous studies. The ideas were further stimulated by specific discussions with senior personnel in two rather different 'service' sectors, namely higher education and financial services (both of which have also experienced high rates of change in recent years).

This chapter examines the connections between organisations, individuals, and physical settings, that influence the effectiveness of the support provided by the workplace. The aspects considered are: working practices, empowerment, effectiveness and efficiency, the position of the individual, and the environmental implications. It builds on the argument by Thompson and Warhurst (1998) that there is no single future work or workplace, in other words, a range of approaches is needed to address variation in location, sector and organisation. This puts the 'new workplace' in perspective, recognising the great complexity in current trends where continuity is as pervasive as change.

Perhaps an appropriate place to start this examination is to reflect on power. Recent concerns to move the perception of FM from an operational to a strategic level can be seen as a desire to increase the status and power of the facilities manager, albeit in a belief that this would be to the common good. George Orwell's definition of power in Nineteen Eighty-Four, a look at the future published fifty years ago, was: "Power is not a means, it is an end. One does not establish a dictatorship in order to safeguard a revolution; one makes the revolution in order to establish the dictatorship." (Orwell, 1949.) By extension, the people in power can define what constitutes knowledge and, ultimately, reality. The location and nature of knowledge and power have tended not to figure highly in FM literature, however. Neither, as Cairns and Beech (1999) have said, has the literature engaged with the debate going on in management circles about the "shift from overt threat to moral coercion". It is interesting that another article published at around the same time observed that although pseudo-power (over the timing and location of work) may be devolved to the workforce, real power (to define and manipulate the knowledge base) remains elsewhere (Myerson 1999). Hence there is a tension between individual freedom and the organisational good, and between flexibility/innovation and organisational control. These tensions are present in much of the debate that follows.

Working Practices

The response to radical and continuing change is a desire for flexibility (to cope with current challenges and adjust to future ones). This can manifest itself in the form of:-

- Contractual flexibility.
- Functional flexibility.
- Temporal flexibility.
- Locational flexibility.

Contractual flexibility will not be considered in any detail here, but it is an important part of enabling more flexible working, and hence will be examined briefly. Research by Guest et al. (1998) concluded that flexible contracts often gave conflicting messages between cost cutting and a high usage of human resources, and could create a climate of unfairness. They perceived the root of the problem to be the fact that H.R. departments were responding to the board room rather than organisational psychologists, and hence contracts were dominated by the demands of labour economists while the subjective experiences of employees were under-emphasised. This dichotomy relates to that concerning efficiency and effectiveness raised later.

Flexible working is often referred to as 'teleworking', which has been defined as: "work based on or facilitated by the network technologies of the information society" (European Commission, 1998). This has been characterised by Huws et al. (1997) as having some advantages (like flexible working hours and new forms of work sharing) which are linked to contractual, temporal and functional flexibility but also others which relate to locational flexibility. Hence they identify five categories:-

- 'Multi-site teleworking' which is partly office and partly home-based.
- 'Tele-homeworking'; working at home for a single employer.
- 'Freelance teleworking' which is also home-based but for multiple employers.

- 'Mobile teleworking' which centres around portable equipment used at a variety of sites.
- 'Relocated back-offices' where work is carried out at a distance from the main premises.

Several studies have tried to quantify the current situation. Compelling evidence of the rate of change comes from Worrall and Cooper (1998) who discuss an Institute of Management/UMIST study that questioned 1,000 managers. Around 60% of these, in the two years under consideration, replied that their organisations had restructured over the preceding 12 months. Further evidence comes from the European Commission (1998) which recorded that the number of employees who worked variable hours from week to week had more than doubled in the previous ten years to over 50% (although some of this may just be small variations due to basic flexi-time systems). The same report also stated that 4 million people 'teleworked' in Europe (2.5% of the workforce), more than twice as many as two years previously. The U.K. was near the top of the league with about 7%, with Denmark and the Netherlands at around 9%. Meanwhile, Dodgson (1998) predicted that in 5 years 10% of work would be done at home, and a study carried out by MORI for Mitel recorded that 36% of SMEs and 59% of Times 1000 companies had teleworking to some degree (Home Office Partnership 1999a).

Hence the picture seems to be one of a growth in flexible working but also of the majority of people continuing to have conventional contracts and a single work location. Furthermore, although flexible working is increasing (with cost cutting and space saving being the major drivers) there are other factors that are working against this trend, e.g. the difficulty of managing different work locations (Industrial Society 1998) and workforce acceptance (Bowers 1997). This resistance is evident from both managers and other employees, and centres around trust, visibility, face-to-face interaction and teamworking, concerns about the erosion of the distinctions between work and home, and the symbolism of place, although it may not be expressed as such. Issues relating to trust and face-to-face interaction will be considered below, while the last two aspects will be discussed later in relation to the individual.

'Trust' for managers often centres on the fear that their teams will not work when away from 'base'. For the team members the worry is that if they are not physically present their managers may not continue to value them. These attitudes are difficult to change because they are embedded in workplace culture, and even the possibility of being 'visible' through new technology is rarely regarded as compensation for the physical separation. The loss of face-to-face contact (anticipated in most types of flexible working) is often perceived as detrimental to communication, it was cited, for instance, as a reason for inhibiting the expansion of homeworking by 62% of employers studied by Dodgson (1998) in work for the RAC. This aspect is important not just in itself but also because it relates to which jobs can be made more 'flexible', and, by extension, which people may exercise choice. An additional factor is that of gender. The proportion of women in low paid jobs is very high, and people in such jobs are likely to accept less favourable 'flexible' arrangements because they have no choice (Huws et al. 1997). Conversely, professional and white collar women tend to be given and/or feel suited to, roles which have a large face-to-face component (Wood 1997) and hence may not be given the opportunity to work flexibly, even though this is what they would prefer. On the other hand employers' beliefs about the preferences of men and women can result in gender stereotyping where women are assumed to be attracted by flexible working arrangements and men by more money. Casey et al. (1997) suggest that such stereotyping "imposes severe constraints on the opportunities open to women and to men who hold different preferences".

Furthermore, the success of 'new ways of working' may also be subject to whether they deliver the culture change that has become associated with them. A 'new breed' of managers has listened to the rhetoric and now, according to the Ashridge Management Index, expects empowerment in a risk taking environment which encourages innovation and has a clear sense of purpose (Home Office Partnership 1999b).

Empowerment

Cairns and Beech (1999) point out that although 'empowerment' has been presented as one of the cornerstones of new ways of working it could be said that 'knowledge' workers always were empowered and hence the crucial question is whether these changes empower the bulk of the workforce. This point was also made by Moon et al. (1998) who comment that most advocates of new ways of working concentrate on the 'knowledge' workers and conveniently ignore the remainder of the workhorse. Duffy (1997) for instance, says that "routine clerical tasks have been automated or exported off-site, away from the creative teams and decision-makers", exported to "where they can be carried out far more cheaply" (the 'relocated back offices' of the Huws classification). Moon et al. (1997) also quote Burdett as saying that empowerment must have authority, responsibility and accountability and that although the term is common the practice is rare. They further use the work of Cressey and Scott to make the point that clearing bank reorganisation had already created, by 1992, advantaged groups who were 'empowered', and others whose work had been debased. The latter were "exposed to more factory-like conditions with their performance closely controlled through computer technology". (This was in the context of over 80,000 jobs being lost in banking between 1990 and 1994.)

The physical manifestation of this last trend is now familiar from call centres in Sunderland to computer programming operations in Bombay. The subject of 'call centres', which usually deal with marketing or customer service help lines, has attracted much attention. The U.K. has half Europe's call centres and it has been predicted that by the end of the year 2000 between 1 in 30-50 jobs in the U.K. will be in call centres. Even if this prediction is exaggerated, it can be seen that call centres are likely to affect a substantial number of people (the growth in on-line interaction may change the nature of the work, but not the basic functions being performed).

Although call centre employees can be termed 'teleworkers' their typically 'disciplined' environments, dress codes, targets and team leader briefings at the start of each shift, link more with 19th century work settings and

Nineteen Eighty-Four than popular notions of flexible working. 'Teleworking' has come to be associated with organisations that are flat, have agreed targets and little bureaucracy. This is the image presented by those organisations referred to above who are in the forefront of the movement, and who consist mainly of creative people ('knowledge' workers). But 'tele-working' is an approach that can also be seen to work in exactly the opposite scenario, where there is a strong hierarchy, work is packaged, exact instructions are given and there is electronic control over progress. This both allows the work to be carried out at any location and uses electronic surveillance to allay managers' fears. Hence organisations in financial services have variously embraced the possibilities of telebanking (cited as being 20%-30% cheaper than conventional banking) and, because of the prevalence of standard forms, the use of 'image and workflow' (where documents are scanned and work distributed and returned electronically). This highlights the tension between flexibility and control mentioned earlier. Dan Preston, a specialist in business ethics at the University of East London, has been quoted as saying "companies aren't really concerned with a little bit of internet misuse or sending a few private e-mails here and there. What they are really concerned with is identifying people they distrust" (Independent 1999a). It can be argued that unless facilities managers engage with such power relationships and ethical issues they are unlikely to make an impact at a strategic level.

It is interesting to compare these developments in the commercial world with those in higher education, where numbers of U.K. students have more than doubled in the last ten years. Academics, it could be said, have always had a system of flexible working, but this has not been true for administrative and technical support staff, nor (officially at least) for many students. In terms of property and facilities many universities still have 'estates' departments, which manage building maintenance and are involved in commissioning new buildings, with other 'facilities' functions distributed amongst a variety of departments. The development of many 'new' universities (former polytechnics) has been characterised by growth, a drive for improved efficiency, and an emphasis on 'audit', in an attempt to demonstrate quality in relation to criteria which are very different from those applied to the polytechnics. This, plus a move to modularisation, has contributed to a change from a relatively flat organisational structure to more centralisation and layers of management (the opposite direction to business). Increasing centralisation may also be observed in some 'old' universities (timetabling for instance), and Gore et al. (1998) found a very strong trend towards greater formality and tightening controls at university level in a study of business schools in 44 higher education (HE) institutions.

Efficiency and Effectiveness

Because of the fairly widespread tendency to use these two terms interchangeably it is perhaps understandable that some commentators have sought to separate them. Duffy (1997) for instance, defines efficiency as "driving down occupancy costs" and effectiveness as "adding value to business performance" and suggests a balance should be struck between the two. An alternative model is that effectiveness encompasses efficiency. Gordon (1999), has suggested four facets to effectiveness; quantity, timeliness, quality, and the ability to deal with multiple priorities. Quantity and timeliness also apply to efficiency. The other two facets, multiple priorities and the perception of 'quality', are likely to vary according to individual and group differences, a theme which has been developed by Garnett (1995) using Multiple Rationality Analysis. The perceived difficulty in measuring these two facets of effectiveness, plus the over-riding importance of (short-term) economics, may explain the influence of efficiency at the expense of effectiveness.

Differences between cultures on the issue of effectiveness and efficiency, and how values are often inextricably linked to built form, are highlighted by Duffy (1997). He suggests that the Northern Europeans have generally preferred effectiveness, while the North American tradition is for efficiency. This is not to argue for crude environmental determinism, rather that the buildings and the culture can be mutually reinforcing. Hence,it could be said that the late 1980s-90s fashion for 'streets' represents a return to tradition in N. Europe, in as much as it recognises the importance of a qualitative aspect, namely chance contact, the opportunity for which is provided by circulation space (seen in the efficiency model as a necessary evil which is to be kept to a minimum).

Space utilisation and space standards are usually considered in terms of efficiency, probably because it is very difficult to measure effective space. The difficulty is not just about quantifying output, but also, as Worthing (1994) and Shove (1993) have pointed out, because what constitutes effective space can change over time and between users. It is important to remember, however, that although space may be dealt with in a quantitative way it will still be invested with symbolic meaning, either reinforced at an institutional level (through allocation according to status) or purely by individual perception. The use of a single mathematical equation of space efficiency to characterise space cannot address the integration of activity, accommodation, perception, and time, that contribute to space effectiveness. These issues have been central to the debate concerning the assessment of buildings and appropriate forms of Post Occupancy Evaluation (Wood et al., 1995). Experience suggests that even when the brief for such an exercise attempts to encompass effectiveness, it is very difficult to ensure that the qualitative and 'stakeholder' aspects have sufficient impact on subsequent decision making. This issue of which aspects are taken into account and whose perceptions are valued is central to the formulation of performance indicators. Ashcroft (1995) warns that, in spite of the way in which they are presented, such quantitative measures are not value-free. With regard to the different priorities of various groups in higher education, Housley (1997) found that property managers tended to see support service functions to faculties as more important than senior managers (usually pro-vice chancellors), while the latter rated managing occupancy costs and communicating value to HEFCE more highly. This suggests a difference in focus between internal service and external justification. Although what constitutes 'value' to HEFCE may be changing, in a recent bidding round for estates funding, criteria used included: the impact on the institution and its students; the coherence with existing strategies; additional benefits (e.g. environmental) and achievements in estate management. In other words, there was an attempt to take some qualitative and 'stakeholder' issues into account.

The move towards 'two tiers' (the 'knowledge' workers and the remainder) discussed earlier, is evident in a slightly different way in H.E. Diminishing government funding in parallel with numbers which continue to increase, mean a recognition that a return to 'elite' universities, or parts of universities is likely. These would be inhabited by mid-career professionals, overseas students and others who can afford to pay, alongside would be a lower level of 'mass' education. The physical embodiment of these two extremes can be seen in recent high profile buildings by well-known architects which create environments similar to prestigious offices and which are in sharp contrast to the wider picture of poor estate condition. The latter is the result of the standard of building during the 1960 - 1970s expansion, followed by years of under investment in facilities and maintenance. In an attempt to tackle the wider problems HEFCE has recently made available 'matching funding' to improve poor estates, amounting to some £250 million over five years (HEFCE 1999]). The implications of this polarisation are very great in terms of raising finance, attracting students, and the conditions in which staff and students work.

The Position of the Individual

There is a tendency in the workplace to refer to groups such as 'marketing', 'finance', etc., who are seen to have a role or particular characteristics in common, but individual variability also needs to be taken into account.

When people who hot-desk are asked what they think about it they usually say something like "I really don't need a desk", rather than "I love it, it's much better than having a dedicated space". It can be suggested that few employees prefer hot-desking to having a fixed desk, rather it is the employers who favour this way of working. The advantages of flexibility for many individuals are real but so are the worries, referred to earlier, especially if flexibility really equates with longer hours as it becomes more difficult to define the boundary between work and home. Home working in particular raises difficulties, not just where the physical or social environments are not conducive for work but also in the more symbolic erosion of temporal and physical boundaries which may cause problems with partners or family members. Even the physical distinctions are diminishing, sofas at work, desks at home, for example.

The symbolism attached to places is a powerful force that works against locational flexibility and particularly hot-desking. This varies, however, between groups and individuals. Goodrich (1986) refers to studies of 13 corporate complexes in the USA from which a picture emerges of personalisation of the workspace differing by job type and gender. Financial personnel and administrators were amongst the people who personalised their spaces the least, while secretaries, and marketing personnel personalised them the most. Women were found to be more likely to personalise the space aesthetically (with plants, posters, and personal items), whereas men tended to display personal achievements or family photographs, with managers and executives using these to enhance their status. One of the companies studied had restricted personalisation in a new building. Here, the researchers found that individuals from groups that personalised their space most described the environment as 'sterile, impersonal, and cold'; while those from groups who personalised the least described it as 'pleasant, comfortable, and attractive'.

People's preferences are influenced not just by their role and gender, however, but also by the values of the wider society (what constitutes 'success', what is fashionable, etc.) and by individual personality type. Goodrich (1986) found that extroverts and introverts, perhaps unsurprisingly, favoured different characteristics of the workplace. More fundamentally, he makes the point that user wants are not the same as user needs, characterising the latter as being deeper and more fundamental, involving the worker in trying to achieve a satisfying and meaningful working life. In other words, he sees needs in terms of Abraham Maslow's famous hierarchy, and argues that these should be addressed and understood in the creation of office environments. The influence of individual variability was also highlighted by Whitley et al. (1995) who found that the personality difference measure 'locus of control' was linked to both job satisfaction and perceptions of physical conditions.

With the breaking down of the distinctions between home and work, work can become the centre of the 'knowledge' worker's life, leaving little time to arrange other aspects. Hence there seems to be a trend for employers to offer services to employees once considered outside their remit, from gyms, through laughter consultants to dating agencies. Some of these require facilities and all need organisation. The intention is to keep employees loyal to the company, but such actions can be interpreted as selling back as benefits those aspects which have been taken away by the consequences of the job (Guardian 1999).

Some current trends (global competition, long working hours, more women working, more people living alone, etc.) suggest a move to a 24 hour society. Huws (1997) and Evans (1998) comment that it is not only most people's personal lives but also much of society that is based on a 'normal' working week - education, pensions, leisure opportunities, etc. Evans (1998) postulates that the logical outcome of a 24 hour lifestyle is a return to more standardised patterns, i.e. shift work. Interestingly, one of the financial services organisations consulted during the preparation of this chapter was considering introducing shift work for its 'core' business, the benefits of increased output and space utilisation being seen to outweigh the implications for individual workers and for cleaning, maintenance and other support services. As Martens et al. (1999) point out, however, what is good for business may not be good for the health of the individual. Although flexible work schedules have smaller effects than other variables, including subjective workload, they contend that it is a matter of extra stress added to pre-existing work-related stress. They also reinforce the point that variations in work patterns affect other aspects of people's lives (e.g. taking part in children's education or leisure activities). As in other studies people who worked changing shifts, or those doing long shifts in a compressed working week, fared worst physically, whilst those on temporary contracts (mainly women) reported significantly more problems relating to psychological performance.

Finally, the emphasis on IT in teleworking assumes not just technical skills but also a particular communication style. There are a variety of styles, however, and by focusing on one there is a danger of excluding people with different skills who may prove important to future teams.

Environmental Implications

Business discourse has tended to ignore physical setting, the term 'environment' typically meaning the economic or social environment rather than the physical one. Recently the 'green' agenda and a growing recognition of the interconnections between different aspects of people's lives have resulted in comments like those of Diane Coyle "Companies exist in actual places, but it is rare to hear about how place can be made to contribute to the productivity of UK plc." (Independent, 1999b.)

In physical terms the present pattern of organisational property is quite often an impediment to new ways of working, and although the perception that teleworking is necessarily linked to a particular space (e.g. the home) is false, the lack of infrastructure, in terms of a local economic development strategy, the provision of local telecentres, and dwelling design are important factors (European Commission, 1998).

The link between teleworking and possible environmental benefits has been acknowledged by a number of sources including the E.C. (Annual Report from the European Commission, 1998), and centres around the reduction in miles travelled. The expected benefits are, for government and society, less pollution and congestion and for employers and employees, a reduction in stress and travel time. The recent Government White Paper, however, (DETR, 1998) recognises some of the complexities outlined below.

The expected benefits are discussed by Dodgson (1998), who quotes his report for the RAC which cites the 1997 National Road Traffic Forecasts of an increase on 1996 levels of: 10.9% by 2002, 20.8% by 2007 and 39.6% by 2017. He then calculates that a conservative estimate of the impact of telecoms would reduce these figures by about one third, while an optimistic estimate would more than halve them. A Norwegian study is also interesting, although comparisons between countries need to take into account that behaviour is influenced by culture. It estimated that if between 10% and 20% of employees in Bergen and Oslo worked at home one day a week the reduction in car traffic would be between 3% and 6% (ETO Homepage 1997). The assumptions built into these calculations are unclear but if they are directly correlated with working hours in an office the predictions should be viewed with caution; humans tend to modify their behaviour to maximise benefit. Hence if rush hour traffic is reduced people who now travel later may change their habits, or public transport users be tempted to use their cars. Some of the people working from home will previously have travelled on public transport, and journeys will be spread more evenly during the day. Which, in turn, will have an effect on provision and on services like season tickets and OAP travel concessions. This kind of strategic switching of behaviour was demonstrated in a study of the Bristol 'park and ride' scheme (Mills, 1997), where it was concluded that although the numbers using the system at peak times looked impressive, when their previous behaviour was taken into account it was found that there was a net environmental loss because of the move away from public transport and the fact that some passengers would not have made the journey at all previously. Furthermore, as the European Commission (1998) points out, some forms of locational flexibility have inherent problems. The replacement of large central offices with smaller dispersed ones used by a nomadic workforce, for instance, is attractive for companies but may be bad for the environment (longer journeys between dispersed offices because people still want face-to-face meetings with colleagues, clients and customers). Flexible working hours also make informal travel sharing arrangements very difficult.

Finally, the DETR/Home Office Partnership study of Cambridgeshire County Council found that staff attitudes to home working correlated with travel time, rather than distance; those living a similar distance away but having shorter journey times were less interested in working at home (Jupp, 1998). Cooke (1998) remarks on the fact that the average journey to work time has remained at just over half an hour since the Victorian era but work distance has increased dramatically. He goes on to comment that improved roads allow employees to make choices like having a longer journey to work for a new job rather than moving house. Hence, the picture is complex, and predictions based on a single parameter are unlikely to be accurate.

Concluding Remarks

One of the points raised at the start of this chapter was the position of the facilities manager. Flexible working implies not just more management but more skilled management, a component of which is a better understanding of effectiveness. This, combined with the fragmentation of property (which means that there may be a reduced but more diverse and flexible portfolio in terms of physical form, tenure and use) could present an opportunity for facilities managers at a strategic level.

The predictions that flexible working will allow organisations to improve the quality of the work experience, the lives of their employees, the environment, and make financial savings need to be approached with caution. Modest changes that adapt 'traditional' patterns have been shown to be popular - working one or two days per week at home, for instance, was favoured by both employees and employers in a recent study (Daniels, 1999). However, such a pattern makes space economies difficult (unless combined with some form of desk-sharing or space leasing), and gains in work effectiveness will not necessarily follow. Furthermore, delivering the expected environmental benefits may not, as we have seen, be as straightforward as at first appears.

More radical moves towards 24 hour work and support service availability have fundamental implications for both society (in terms of organisational framework and physical infrastructure) and individuals (in relation to their health, relationships and socialisation). A special feature on the occasion of the Harvard Business Review's 75th anniversary looked at implications for the future, with contributions from such luminaries as Peter Drucker and Charles Handy (Drucker et al. 1997). A strong theme emerged of the importance of community, and, even if the driver is organisational preservation rather than philanthropy, this orientation is interesting. The associated editorial comment that the challenges identified by the gurus are more cultural than technical or rational, reinforces the value of the European preference for effectiveness.

Hence current trends may be seen as both positive and negative, and the political and social climate of today ambivalent enough for at least some of the future imagined in Nineteen Eighty-Four to remain a possibility to be guarded against.

This chapter has ranged widely in trying to map the ways in which property and facilities support the world of work, in order to highlight some of the key considerations that may influence the future. What emerges above all is a need to remain open minded, to apply analytical thought, and to embrace continuing possibilities: "The key to harnessing more of the commitment and potential of people is to allow them to work and learn in whatever ways suit their needs and preferences, match their natural aptitudes and otherwise allow them to be more effective...The flexible organisation may not need standard ways of working, standard approaches to learning, or standard technologies." Coulson-Thomas (1997, p.8).

References

Ashcroft K (1995) *The Lecturer's Guide to Quality and Standards in Colleges and Universities*, The Falmer Press.

Bowers C (1997) *Optimising the use of Office Space: Hot Desking Opportunities*, University of the West of England, unpublished MSc Facilities Management dissertation.

Cairns G. and Beech, N (1999) Flexible Working: organisational liberation or individual straight-jacket? *Facilities*, Vol 17 No1/2 Jan/Feb, pp.18-23.

Casey B. Metcalf H. and Millward N. (1997) *Employers' Use of Flexible Labour*, Policy Studies Institute.

Cooke P. (1998) Integrated Transport? In: *Is Commuting to Work a Necessary Evil?* (Ed. by T. Osborn-Jones) Future Work Forum Report, Henley Management College.

Coulson-Thomas (1997) The Future of the Organisation: Electronic Commerce and Corporate Transformation, in Frost Y., *Managing Work Remotely*, Future Work Forum Report, Henley Management College.

DETR (1998) *A New Deal for Transport; Better for Everyone Government White Paper on the Future of Transport*, HMSO, UK.

Daniels I. (1999) *The Role of Employers in Reducing Car Use*, University of the West of England, unpublished M.Phil thesis.

Dodgson J. (1998) Motors or Modems? In: *Is Commuting to Work a Necessary Evil?* (Ed. by T. Osborn-Jones) Future Work Forum Report, Henley Management College.

Drucker P. Handy C. and Dyson E. (1997) Looking Ahead: Implications of the Present, *Harvard Business Review,* No 20, September/October pp.18-28.

Duffy F. (1997) *The New Office*, Conran Octopus, London.

ETO Homepage (1997) *Can Telework Reduce Traffic Congestion?* European Commission Directorate General XIII, http://www.eto.org.uk/faq/faqtpt1.htm.

European Commission, The (1998) *Status Report on European Telework; 'Telework 1998'*, Directorate General XIII, September.

Evans C. (1998) *Managing the Flexible Workforce: Enhancing Performance by Aligning Organisational and Individual Needs*, Roffey Park Management Institute, February.

Garnett D. (1995) Multiple Rationality Analysis: An approach to reconciling the competing values and interests associated with housing renewal schemes *CASTLE/IFS Conference, Sustainable Development; Counting the cost, maximising the value*, Harare, Zimbabwe, August.

Goodrich R. (1986) The Perceived Office: The office environment as experienced by its users. In: *Behavioural Issues: Office Design*, (Ed by Wineman) Van Nostrand Reinhold, New York. pp.109-133.

Gordon G. E. (1999) *The Last Word on Productivity and Telecommuting* http://www.gilgordan.com/downloads/productivity.txt.

Gore Chris, Steven, Valerie and Bailey, Mark (1998) Analysis of the Effect of External Change on the Management of Business Schools Within the Higher Education Sector, *Total Quality Management,* Vol. 9, Nos 2/3 pp. 249-258.

Guardian (1999) *Love me Love my Job*, 21 April.

Guest D. Mackenzie D. and Smewing C. (1998) *Innovative Employment Contracts: A flexible friend*, Working Paper OP 199, Dept. of Organisational Psychology, Birkbeck College, University of London.

HEFCE (1999) Council Briefing No 22, March, *Higher Education Funding Council for England.*

Home Office Partnership (1999a) Flexibility http://www.flexibility.co.uk/contents.htm.

Home Office Partnership (1999b) Flexibility http://www.flexibility.co.uk/ashridge.htm.

Housley J. (1997) Managing the Estate in Higher Education Establishments *Facilities*, Vol.15 Nos 3/4 March/April pp.72-83.

Huws U. et al. (1997) *Teleworking: Guidelines for Good Practice*, Institute for Employment Studies Report 329.

Independent (1999a) *Your Office Manager is Watching You*, 17 January

Independent (1999b) (Enterprise Issues) *All that Clusters Could be Gold-for Britain PLC*, 14 April.

Industrial Society, The (1998) *Managing Best Practice 46; Flexible Work Patterns,* April, The Industrial Society.

Jupp S. (1998) Telework and Transport In: *Is Commuting to Work a Necessary Evil?* (Ed by T. Osborn-Jones) Future Work Forum Report, Henley Management College.

Martens M.F.J, et al. (1999) Flexible Work Schedules and Mental and Physical Health. A Study of a Working Population with Non-Traditional Working Hours, *Journal of Organisational Behaviour,* 20, pp. 35-46.

Mills Geoff (1997) Bus Based Park and Ride: Towards Sustainability? In:

Evaluating Local Environmental Policy, (Ed. by S. Farthing) Avebury Studies in Green Research.

Moon C. and Swaffin-Smith, C. (1998) Total Quality Management and New Patterns of Work: Is there life beyond empowerment? *Total Quality Management,* Vol.9, Nos. 2/3, pp.301-310.

Myerson J. (1999) Real People, *Premises and Facilities Management*, January, p.19.

Orwell,G. (1949) *Nineteen Eighty-Four.* 1954 Edition cited, Penguin. UK.

Shove E. (1993) *The Black Holes of Space Economics*, Buildings and Society Research Unit, University of Sunderland, January.

Thompson P. and Warhurst C. (1998) Hands, Hearts and Minds; In: *Workplaces of the Future,* Macmillan. (Ed. by P. Thompson, and C Warhurst.)

Whitley, T D.R. et al. (1995) The Environment, Comfort and Productivity: The Role of Individual Differences Including Locus of Control, paper presented to the *Healthy Buildings '95 Conference.*

Wood, S., Worthing D. and Hobbs P., (1995) An International Typology of Post Occupancy Evaluations, *Proceedings of the RICS Cutting Edge Conference 1995*, RICS.

Wood S. (1997) Unpublished paper, University of the West of England.

Worrall L. and Cooper, C.L. (1998) *The Quality of Working Life 1998*, Survey of Manager's Changing Experiences Institute of Management and UMIST

Worthing D. (1994) Strategic Property Management, In: *CIOB Handbook of Facilities Management,* (Ed. by A Spedding) Longman Scientific & Technical.

11 Self Selected Workplaces

Richard Watts, RKW Space Management Consultants

The re-engineering of work and the goal of increasing productivity to max-imise profits are likely to be continuing themes of the new millennium era. As ways of doing business become ever more diverse, so too will the types of work settings that are appropriate for business. Information based organisa-tions, measuring their performance on non-linear and intellectual activities, have never concerned themselves with the traditional systems of space bud-geting and allocation. There is emphasis on working towards pre-defined goals, with little dependence on job descriptions or management structure. Key staff are given more freedom to choose when, where and how they get the tasks completed. As this model is seen to work, it is likely to become the way of business for an increasing proportion of progressive enterprises.

Change is occurring in the supply side too. As more organisations begin to move their real estate off the balance sheet, new opportunities for greater choice and variety in the workplace become viable. The explosive growth in the serviced office sector has already shown that there is demand for much more flexibility in commitment and availability than the traditional office marketplace can offer. In a recent survey there were over 36 separate serviced office locations in London alone. Not only are suppliers, such as HQ and Regus, building more business centres, but they are offering a wider range of products, ranging from a 'virtual office' messaging service, up to a complete custom-built office fit-out. Currently their marketing tends to be directed towards multi-national organisations who require a presence in a spread of geographic locations. It seems likely that they, or new competitors, will expand to appeal to a wider range of enterprises and offer a choice of quality standards and pricing options.

The hotel industry has moved towards providing individual work space, in the form of 'wired' hotel bedrooms, as well as a range of conference facilities. A few have expanded their range of offerings by providing purpose-built busi-ness centres and workspaces. For people on the move, motorway service areas already are frequently used as meeting and work places for business people on the move, though little is provided to meet their particular needs. The business centres at most airports offer, by contrast, a quality, well-equipped work setting. In this field, we should expect to find new players offering new short-term occupancy spaces and pay-as-you-work facilities in direct competition with the established workspace providers. With the growth of shopping as a leisure activity, one might soon expect to find business cen-tres within large out-of-town malls.

We can also anticipate a increasing range of business service offerings, covering virtually every non-core business activity. Mail and call handling, conference services, communications and IT are just some of the services currently available on an outsourced short-term basis. None of these are dependent on the client's physical location. As the trend towards more outsourced services continues, competition will ensure the availability of these types of services to smaller companies and at more attractive rates. Much has been written about the impact of the Internet on commerce - suffice it to say that there are well publicised examples of location independent or 'virtual' businesses growing at an exponential rate.

The key trends of IT convergence and performance will offer ever greater time and space options to end users. The first electronic calculators were so expensive that they were normally purchased by businesses and shared amongst users. Today, they are bought by individuals for less than £5 or even given away. If PCs and mobiles follow this price/performance curve, then we can expect that users will be willing to buy whatever suits them best. Assuming that the current incompatibility problems will be solved, any voice or data service will be instantly accessible to any authorised user from any location at any time. In consequence all the desk top services can effectively be 'outsourced' to end users to self-select and self-connect. User teams then can reconfigure computing power to do the jobs they want it to do, in support of their own work with remote IT infrastructure. At every level, new cordless technologies - WANs, LANs and now PANS (Personal Area Networks) are starting to release workers from the tie of phone and data cables to any fixed location.

Recent surveys of work groups have shown that end users are enthusiastic about having choice in the selection of their work setting. The ability to choose where to work on a dynamic basis is very popular, but more difficult to manage than the more traditional structured approach. In future, as businesses need to compete through continuous improvement, their priority will be to encourage employee interaction for invention and product innovation. In this scenario, work activities are likely to be divided into projects, each of which has a manager, charged with a clear mission and empowered to get all the resources necessary to achieve the result. Space becomes simply one of a number of administrative issues which need to be resolved in order to progress with the project's main objectives. As every facilities manager knows, the block to such an approach has, up to now, been the long lead time to acquire space, not the ability of entrepreneurial managers to specify what they want.

Managing such groups will be a key challenge for the businesses of tomorrow. Management has to be oriented towards setting out the aims and results required from an individual's work, within set parameters of time and budget. This calls for a relationship of trust and openness. With the increasing globalisation of work, managing remote staff is now common in many multinational enterprises, but remains a source of concern to many traditional UK managers. Confidence in managing such dispersed staff will become a key factor in the broader acceptance by management of location independent working. This will require a systematic training programme for managers and for co-workers. Techniques for objective setting, performance measurement and staff motivation are established components of most management education today. No new revolutionary concepts need be invented in order to manage nomadic workforces but perhaps more innovation in the delivery of management education and an increasing realisation of the relevance of such tools to the new world of location independent working.

Bringing these threads together, the entire workplace selection process can be devolved to an individual user level. Armed with only a smart credit card individuals will be able to select the work setting and services they require on a pay-as-you-use basis. An increasing number of options will exist, none of which result in a residual or long term commitment. A menu might include the following types of work settings:

Home working: an attractive option for many people for some part of the working week. Recent surveys have indicated high levels of satisfaction for homeworkers to complete administrative tasks and concentrated study. However, home working for more than about 50% of the working week can result in isolation and loss of team spirit. Given a real choice, home workers usually recognise the need for a balance of time spent in concentrated productive isolated work and the need to interface with co-workers and decision makers. In this scenario, the choice can be theirs to make.

Local teleworking centres: quality access to remote databases and applications whilst on the move is essential to many workers and often inhibits location independent working. Convergence of technologies and price performance are making it possible to interact with corporate systems without direct connection to high speed data lines. This will in time supersede the need for local workplaces, such as telecentres or local drop-in centres, which in the short term can offer higher quality connectivity and security, plus a range of business services and facilities for the mobile worker.

Customer or supplier based working: many business today have groups of contract staff undertaking defined tasks (e.g. writing software) housed in the client's premises at no apparent cost to the contractor. This will only work as long as the customers have premises with surplus space to accommodate third parties. In a self select space scenario, it seems likely that service providers will need to include the cost of their housing within the charge they make for the service provided. This will increase the pressure to make space and facilities available for the duration of a contract, but without any residual commitment.

Business centre based working: work is undertaken from a ready equipped business centre, either purpose-built or as part of a hotel, leisure or transport centre. Space and services are normally offered on a pay-as-you-use basis with no residual costs.

Redundant space: empty buildings can sometimes be acquired without long term commitment and adapted by a team of people to suit their needs. The fitting up could be leased, rentalised or written off over the predicted life of a project.

Portable space: the comfort and functionality of modern portable buildings, ranging from linkable site cabins to fully fitted mobile offices, offer a further range of possible work settings. Already established within the construction and film production industries for example, will the portable workplace become an increasingly common sight in the car parks of many other types of enterprise?

Unlike today, the choice of work settings would not be so influenced by the need to maximise the use of existing corporate real estate, because there may very little left. Any requirement for a prestige headquarters would have to be self justifying on the grounds of a marketing or branding need. Even this does not have to be owned or leased in the traditional manner. For example, Regus already provide Anderson Consulting with a complete 'packaged' headquarters in Holland. Regus provided the building and the infrastructure on a price-per person basis with little or no capital investment expected from the client occupier.

Of course, there will be corporate structures, both physical and organisational. It is vital to retain a core which houses the 'crown jewels' of any organisation, without which there is no mission or purpose. As Charles Handy has pointed out, if there is no common goal, people will put their own goals first. The expression of this key purpose may require corporate buildings. Technology centres may also be retained as a central resource - they will be the peopleless offices of the new millennium. Certain corporate activities can best be undertaken in one fixed location, but the many functions will find a new freedom.

The type of work setting which groups of users might select will be different from the traditional office model. Managers of large scale cross-functional projects feel that, whilst location will remain the first priority, other aspects of standard corporate workplace are not important to them. They are looking for more freedom, fluidity and excitement that is available in current work settings. Portable, informal furniture, re-configurable team space, transmutable systems in lively stimulating environments are required, not corporate universality. The actual cost of housing people is simply part of the cost of undertaking the project and will be approached differently by different project managers. A pay-as-you-use approach offers creative project managers the opportunity to create a work setting based on their own vision of an ideal team environment.

In California, creative small businesses are eschewing conventional office shells and seeking old sheds and warehouses with huge floorplates and high ceilings to create team spaces. Some current examples of this DIY approach to workplace design may not look very attractive, but are the products of creative and innovative minds. The wiring may be unstructured and the fitting-up supplied by the local handyman, but the inhabitants have created a stimulating work area in their own style. They devise new metaphors to describe their work settings - hives, dens, prairies, porches - which replace the connotations of a conventional office.

A logical extrapolation of office space management in an uncertain world is for enterprises to hand over total responsibility for addressing and solving space demand to their staff. Employees could be reimbursed, either individually or in work groups, with the expense of providing themselves with all the facilities and space required to undertake the allotted tasks. This might include loans to enable homes or other premises to be set up for office activities, as well as the reimbursement of the expense of using any combination of the 'office hotelling' options described earlier. The market place will develop different service levels and pricing structures to meet a range of user needs, perhaps on a star grading system.

Good communication of every sort is essential if out of sight is not to become out of mind. Whilst E-mail and voice mail video are the backbone of informal business communication today, interaction between distributed members of the group will increasingly be via collaborative software, which allows asynchronous or 'time shift' conferencing. This type of software allows issues to be discussed and information shared where the participants have different working patterns. Place this type of application on the company Intranet and it becomes universally available and cost effective. Intranets are already being used on a global scale by companies such as IBM to offer a self-service reservation system for a range of facility services. Cisco is starting to use its Intranet based enterprise-wide calendar system to allow employees to self-select types of company events which are relevant to themselves and be advised when they are scheduled to occur.

Video conference equipment that can be attached to the PC or hired on a pay-as-you-use basis, will be available to any working team. Face-to-face meetings will not however be superseded. Shared concerns to avoid individual isolation and encourage team building will legitimise the human need for regular face-to-face contact. However, this type of activity could be located in any stimulating environment which is conveniently situated for the participants. There are already unusual locations which can offer a combination of a conference setting within a recreational or leisure element, such as museums, stately homes and theme parks. Short term pre-booking may be required but again there is no long term commitment to occupying all the space or using all the available facilities.

A risk in allowing end users to acquire space and facilities is that, as amateurs in the property marketplace, they are on unfamiliar ground. The user's empowerment would have to be carefully limited to ensure that only short term commitments are permitted. A credit limit on the smart card would seem to be a simple way to control this. Users may need help to find their way in this new marketplace. One can speculate on potential opportunities for work-setting consultants, guiding users through an ever expanding range of options and advising them on the best buy!

'A Warehouse' Los Angeles Ca. (1996) illustrated in *Eciffo*, Autumn.

Gleeson A. (1998) A guide to temporary offices in London, *Flexible Working*, May.

Handy C. (1995) *Beyond Certainty*, Hutchinson, UK.

Hoare S. (1999) The slicing on the cake, *FM World,* Jan/Feb.

Laing, Duffy et al. (1999) *New environments for working*, . BRE/DEGW,
McLennan P. & Cassels S. (1998) New ways of working: freedom is not compulsory, *BIFM Conference*, Cambridge.
Scase R. (1998) Keynote address, *BIFM Conference*, Cambridge.
Watts R. (1999) Unpublished POE for Cellnet.

12 Workplace Democracy

Robert Grimshaw and David Garnett,
University of the West of England

An increasing percentage of the working population in both the UK and USA are employed in offices and Kleeman (1986) has demonstrated that the impact of office environments on the well being of individuals is of major concern. There is a growing body of evidence from the behavioural sciences of the nature of the link between the individual and the workplace, in terms of the harmful effects of adverse environments and the potential for supportive environments to improve organisational performance. Taken as a whole the evidence is compelling; it supports the view that participation in the design and management of office space by users, as a mechanism to combat the negative impact of office space, could be of both social and economic benefit.

Studies by Seligman (1974) and Glass and Singer (1972) support the view embraced by many social psychologists that there is a link between personal control mechanisms in the environment and individual well being. Averill (1973) has demonstrated that when personal control is lacking, people are less happy, less able to make decisions, more apathetic and decreased productivity follows. Studies into institutional environments, typified by Goffman (1961), recognise these reactions as early signs of institutionalisation. Steele (1973) has also argued that cognitive control follows from experience of choice in the physical environment, leading to individual ownership of environmental outcomes.

In the design disciplines, Lang (1987) has demonstrated the links between the design disciplines and environmental psychology, and both Gibson (1979) and Barker (1960) have put forward mechanisms to evaluate interactions between people and the environment. Support for these links also comes from Wineman (1986), Vischer (1989), Sundstrom (1984) and Preiser et al. (1991). More recently, the stress generated by poor quality environments has been investigated by Moore and McCoy (1998) who state that "building design has the potential to cause stress and eventually affect human health. ...Stimulation, coherence, affordance, control and restoration are a preliminary set of environmental dimensions inter-related to stress".

One area of user need that has received particular attention is privacy. From Westin's (1970) and Altman's (1976) early work on privacy as a fundamental access control mechanism that protects the self, more studies have followed which apply specifically to the workplace. These are consistent in supporting privacy as an important aspect of psychological comfort and control. Pedersen (1997) asserts that "too much or too little interaction compared to the optimum desired is unsatisfactory" and that the "process of privacy regulation is salient for the person". Kupritz (1998) in her study of engineers at Gulfstream, reported that "the engineers rank several items associated with privacy regulation immediately after the four design items that are seen as necessary to perform basic job functions". Both support the view that privacy has an important role to play in the workplace.

However, a key element that is missing from the debate is a general acceptance by organisations that the physical environment is anything but neutral. Donald (1994) reports "there has been a lack of research that has focused on the management processes and beliefs that help to shape and structure the environment". This lack of research has restricted the debate about the relationship between the design and management of office environments and organisational efficiency. In particular there has been little debate on either: -

- The workplace as a facilitator of work and its contribution to the success of the organisation or
- The role of employees, either individually or as groups, contributing to the shaping of the work environment.

Barriers to the Acceptance of User Needs

Traditional and entrenched barriers to the relevance of user needs or user participation in design and management of the workplace are apparent in both the design professions and business. Lang (1987) has traced the uneasy relationship between behavioural sciences and the design disciplines. This analysis reveals a fundamental gap between the 'arts based' design disciplines and the behavioural sciences, with a fear that science will reduce the creative aspects of architecture. Many designers avoid the behavioural sciences on the grounds "that their creative roles will be diminished if they rely on the sciences". This is explained by Phillips (1996) in his analysis of the impact of architectural psychology on the design professions. He states that "in almost no cases have the predictions resulting from the work of 'dry' architectural psychologists had practical value for architects, and while the work of wets might have led to greater understanding of, and empathy for, users, such understanding seems to have done little to inform the practice of architects". Phillips even casts doubt on the value of the one practical outcome of research, the P.O.E. techniques promoted by Preiser et al (1988). He reports that "P.O.E. studies have made virtually no inroads into the practical work of private architects". This is symptomatic of the general failure of the dialogue. This does not invalidate the effort but illustrates the cultural barriers that need to be overcome.

Barriers to acceptance of user needs in the business community are equally rigid. Sundstrom (1984) in tracing the history of the office over the last century, clearly shows that Taylorism is firmly entrenched. In particular, the design of work tasks and the allocation of workspace have typically been part of a top down, autocratic system that reduces employee involvement and responsibility to a minimum. This 'classical' model of the firm is driven by an unwillingness of management to give up control and a reluctance by employees to take responsibility. Argyris (1998) reports that, in spite of the push for empowerment in large organisations in the last 30 years, in reality little has changed. Baldry et al. (1998) trace the continuation of Taylorism into new working environments like call centres, under the evocative label of 'Bright Satanic Offices'. They also report that while the organisations in their study were looking for "flexible space that would facilitate IT use around team working" there was "a strong case for cost considerations, driven by short term accountancy needs, to take precedence".

Rational for Change

But perhaps the first cracks are appearing. The 1999 Royal Society of Arts lecture on 'Ethics and the Role of Business in Society' may be symptomatic of a shift in the general thrust of business away from its concentration on profits and share price, where obligations to employees and customers alike are incidental to the primary duty to the shareholder. John Kay (1999) stated that his research demonstrates that "successful businesses serve the needs of their customers, provide a rewarding environment for their workers, satisfy the needs of those who finance them and support the development of their communities" and claims that to sustain long-term development, businesses need to be managed for all their stakeholders. This validates the rationale for user/employee involvement. Grimshaw (1999) has shown that many previously stable social and economic relationships are altering under the relentless progress of IT. Those that affect the workplace and the relationship between organisations, their employees and the physical environment, are particularly relevant to facility management. Although generalisations must be treated with caution, three trends can be taken as of particular interest.

1. **Flexible Working:** the growth of flexible working has been rapid. Apgar (1998), for example, reports that "30 to 40 million people in the USA are now either telecommuters or home-based workers", This has major implications for many aspects of the relationship between employer and employee, including the control and support of off-site working, more responsive non-dedicated workplaces and the maintenance of effective communication. Parker et al. (1998) identify several key risks for individual employees when introducing more flexible patterns of work including "lower control over tasks; intensification of work; insecurity; conflict over new work roles; a sense of unfairness and violation; and a lack of appropriate skills". Participation in the design of work tasks (and by implication the environment that supports them) is seen as a way of mitigating these effects. Parker suggests that the positive benefits of participation include enabling employees to design more appropriate jobs; promoting ownership and understanding; enhancing perceptions of fairness and developing appropriate skills. Positive benefits for the organisation are also claimed including utilising employee expertise and understanding the necessity of employees being self-managing.

2. **Team Working:** Osterman (1994) has noted the growth of semi-autonomous work groups or 'self-directed teams' in both the USA and UK. The benefits claimed are manifold but the implementation of team working often fails to take into account the social dynamics of change (Dunphy and Bryant 1996) and benefits are often patchy. Parker et al.'s (1998) analysis of the factors that influence the success of team working

point to the conclusion that the benefits of team working, both in terms of individual and team productivity, are strongly correlated with the amount of participation by all team members in work design. They conclude that whilst a self directed team can have benefits for both individuals and organisations, "a strategy of partial employee involvement in change will lead to team working only having partial employee and organisational benefits". This implies that only full participation will be effective in promoting team working. It is a short step from general employee participation in the design of work tasks, to argue that they should also be involved in the design and management of the workplace itself .

3. **Knowledge Management:** the rapid development of this area of study is indicative of a major change in how organisations perceive their core business. There is growing support for the view that core business is less about product and more about the knowledge that creates the product (Murray 1999). In this situation the generation of knowledge and its management through the whole business process becomes a key organisational issue. The creativity of individuals and groups must be facilitated both by changes in management style and the provision of stimulating physical environments. Further, Murray argues that the free flow of knowledge around the organisation via both formal and informal communication networks is the key to unlocking and utilising that knowledge. This places an increased emphasis on how workplace design facilitates both formal and informal networks. The former is under the control of the organisation itself but the latter can only be elicited via the participation of employees.

Democracy and the Workplace

It is clear that powerful arguments can be put forward to support user participation in many aspects of work including workplace design and management. Participation implies both rights and responsibilities that can be compared to the democratic rights of citizens. Democracy can be defined as the 'the practice or spirit of social equality' and embodies the rights of individual citizens to participate. However, it has flourished as a political system not because of social equality, but because it has consistently generated a superior economic performance to competing political systems. The participation of citizens in decision-making has allowed democratic systems to be much more responsive to change and avoid the inflexibility that has been characteristic of command economies. Is the same true of organisational cultures? If it is the significance of participation to business is related less to social considerations than to sound economics.

Democracy is about the accountability of the state to its citizens and the legitimisation of its authority in civic life. From the earliest, physical space has played a symbolic role in democracy; the Greek ideal of 'demos' depended upon the provision of a clearly identified physical place where ideas and opinions could be continuously exchanged in an open and free manner. This use of space enabled the enfranchised populace to act as a political unit and make decisions that had collective approval: all citizens participated. The shift from city to national state necessitated a parallel shift from participative (direct) to representative (indirect) democracy where the meeting places of the elected took on great significance as symbols of the working of collective government.

However, citizen rights demand constant vigilance. The current concerns about the health of democracy, typified by the stakeholder debate, show there is growing awareness that real power is defuse; many key policy decisions are not being made by elected representatives within Parliament but by other groups in other places. Decisions that impact on civil life are increasingly being made by informal associations of politically influential individuals, professional lobbyists, appointed quangos, and deregulated commercial organisations; public access to these decisions are limited or non-existent. This threat to the functioning of modern representative democracy is leading to the development of new techniques for citizen participation in a attempt to widen the scope of public involvement in key policy decisions (Renn et al. 1995) The object is to widen the 'decision constituency' and ease the tensions that exist between representative and true participative democracy. Renn also refers to the growing literature on governance and public participation in conflict resolution.

Although the workplace is not a political system the benefits of an inclusive approach need to be explored. 'Democracy' in this context is connected less with equality than participation. For the workplace and its 'decision constituency', the participation of all stakeholders, with the knowledge and experience they bring to the organisation, is desirable to sustain economic success; the well being of stakeholders is an important aspect of that success. The relationship between power and responsibility in an organisation affects its ability to respond; short-term expediency can lead to long term atrophy. Participation is beneficial to an organisation's long term flexibility and success, especially as the shift towards different and looser methods of working gathers pace.

The Design and Management of Shared Space

If a participative 'democratic' workplace is to be promoted, the main design challenge will be where different groups of employees share the same space. However, research in this and other sectors has demonstrated the problems of designing and managing shared space, especially where different groups have legitimate rights to use the same space. Steele (1973) has identified control over the use of space as a key issue that closely reflects the power relationships within the organisational culture. Perceptions of control have to be recognised and accommodated if groups are to feel equally comfortable and empowered in shared space. Grimshaw (1995), looking at supported housing where shared social space plays a crucial role in developing life skills, concluded that shared space depended for its success upon designers understanding the legitimate needs and concerns of each group and clarity around mechanisms for the appropriation of space. Failure to take account of these needs through a lack of participation in the design process, led to problems built into the physical structure that could only be alleviated through often stringent management policies. The result was that shared social space was either overcontrolled or access so restricted that it became effectively redundant. Few successful examples of functioning shared social space were found. The mechanisms for capturing knowledge about user needs were inefficient and participation by any of the parties concerned was very limited. These findings have direct relevance to the way shared workspaces are designed and managed.

Relevance to Facility Management

The issue of user participation in the design and management process is important to the development of facilities management. Its pioneers anticipated a role that could co-ordinate the partnership between organisations and their employees at time of rapid change in the workplace. This was driven less by humanitarian concerns than by economics.

> "Viewing the briefing process as a form of organisational development shifts facility planning and design issues from a narrow to a broad focus; from a concern for the physical environment per se, to a concern for how the planning and design of facilities can contribute to employees motivation and commitment; clarification of corporate goals and objectives; analysis of role boundaries and relationships; and rethinking of strategy and tactics. All of these are a form of organisational development that can affect an organisation's ability to meet its objectives".
> (Becker, 1991).

Becker's views on participation and positive organisational outcomes have been marginalised as facility management has moved more towards a broader role in organisational support. There are signs that this movement is leading to the erection of barriers within FM similar to those previously described for the design professions and business. Perhaps the initial concern for user needs is being forgotten. In this respect Donald (1994), writing as an environmental psychologist, provides a unique view of the development of the FM profession.

His study of the factors behind the degeneration of new office facilities in large organisations in the first two or three years after either occupation of a new facility or refurbishment of an existing building, identified two pressures for environmental change. These were extrinsic pressures which were generated by senior management, usually large scale and imposed from the top, and intrinsic pressures generated by individual employees and groups of employees, which were usually small scale and communicated in a bottom up manner. Whilst it would seem necessary and desirable for facilities managers to respond to both pressures for change, Donald reports a very different picture. Facilities managers were seen to respond only to extrinsic pressures generated by senior management and not to the intrinsic pressures generated by employees. Further he states that "the role of FM in relation to intrinsic pressure seemed to one of preventing or inhibiting such change rather than facilitating it". He paints a bleak picture of facilities managers with little appreciation of either the dynamics of the task-based workplace or the psychological needs of individual employees.

If this study is typical, it supports a growing concern that facility managers see themselves as responsible only to senior management and that they have abandoned any attempt to act as advocates of the needs of users. The early hope that they would become proactive facilitators of organisational change seems to have been dashed. As described by Donald, facility management is a collusive role that reinforces the traditional conflict model of employer/employee relations. A typical quote from a facility manager in the study says it all: "If freedom to make changes were given to the workers then anarchy or chaos would break out".

Donald's identification of conflict, subversion and apathy as the response of this approach by employees is remarkably similar to the response of those who experience institutional environment as reported by Goffman (1961). To these Evans and McCoy (1998) would add a fourth: stress generated by the lack of control and ability to influence the immediate environment. This raises important issues for facility management. It cannot promote policies that result in conflict, apathy, subversion or stress, especially when organisations need to facilitate change and promote innovation. If the long-term health of an organisation depends upon participation by employees, then FM must be prepared to challenge orthodox management beliefs and if FM is to be recognised as a 'profession' it cannot be seen purely as a blind channel for traditional management policy.

Methodological Issues

Conceptual frameworks for addressing the issue of user participation already exist. In architecture, Lang's model of positive theory for design (1987) presents a process that starts with the impact of design on the user. In business Argyris (1998) puts forward an 'External Commitment' model for empowerment based redefining power/responsibility relationships within organisations. However, if rewards for participation are to be judged in terms of economic performance, then a methodological approach based on economic theory needs to be explored.

Because space is a fixed and limited resource, a consequential opportunity cost is automatically imposed whenever a decision is made to utilise an internal area for a particular purpose. Current work at the University of the West of England's Construction and Property Research Centre is seeking to develop the concept of Performance Related Space that dynamically integrates space planning and utilisation. It provides an approach to resource allocation that more consciously relates the use of internal space to the corporate mission. Although the proposition that 'effective space use is performance related' is axiomatic, initial findings indicate that the criteria used for space optimisation in offices are too restricted in terms of the range of cost and output criteria. The narrow focus on short-term cost is counterproductive. In this context, it can be argued that the cost-output consequences of creating shared work areas are seldom subjected to a systematic analysis. Performance Related Space is grounded in concepts taken from the emerging literature on Multiple Criteria Analysis (MCA) and Multiple Criteria Evaluation (MCE). This literature points to a growing demand for a more inclusive approach to decision making. Such an approach seeks to develop evaluation criteria for a fair and competent decision discourse where fairness is defined as the sufficiency of opportunities that participants have to put forward their opinions and to protect their interests, and competence is related to the adequacy of the analytical tools. (Renn et al. 1995: Quinn, 1988)

However, Performance Related Space cannot develop as a concept without consideration of the issues raised above. The analogy of the role of democracy as a mechanism to facilitate development and equate power and responsibility in society as a whole offers a framework for debate. MCA and MCE appear to offer a methodology grounded in economics but which takes account of both technical and socio-psychological considerations. This leads to a model of the democratic workplace that encompasses the following issues:-

- Offices that are seen as places and not spaces with all the attendant implications for symbolism related to belonging and identity.
- Offices that are not designed around static dedicated workstations but have a configuration based on real patterns of work, social interactions, knowledge flows and the need for communication. This must include off-site support.
- A visible negotiation process for both the design process and subsequent management of the workplace with the participation of all relevant parties or their representatives. This must take into account the legitimate needs of all parties and allow for negotiated compromises where necessary.
- All parties must be able to see clear benefits and responsibilities.
- Solutions should evolve organically and not be bound by rigid conceptions of generic approaches.

This is in line with Preiser's assertion in his Manifesto for Humane Environments (1991) that "the process of environmental design must directly involve those affected by the outcome. User participation at some level of design decision making and users planning involvement will achieve more satisfactory solutions if properly managed".

Conclusions

The concept of a participative 'democratic' workplace is worthy of consideration because it responds to emerging imperatives in both business and facility management. The meaningful participation of users/employees in the design, configuration and management of office space has tangible benefits for both individuals, groups and businesses, and reflects changing relationships within organisations.

However, the barriers to recognising the relevance of user participation that are entrenched in business, the design professions and facility management must be overcome. A participative approach challenges the classical conflict model of employer/employee relationships but still needs substantial empirical research to test and validate the concept. Facility management should play a role in this process. It can mediate between the demands of the organisation and the needs of users during both design and management phases of workplace provision and ensure the participation of all parties.

The importance of this process to the professional development of facility management is clear. It will help facility management to regain its concern for the relationship between organisations, employees and physical space; this may be the key to organisational development and is driven by sound economic principles. In 1991 Becker saw employee participation as a key FM strategy, stating that "because staff are the greatest cost (and resource) to any organisation, involving them makes sense both for the better solutions that are likely to emerge and the greater commitment to decisions that are made". In a rapidly changing world 'commitment to decisions' will be an important aspect of long term success and promoting a 'democratic' approach to workplace design and management is an important element. As Agpar (1998) reflects, it "offers a profound opportunity to benefit both the individual and the enterprise".

References

Altman I. (1976) Privacy: A conceptual analysis. *Environment and Behaviour*, Vol. 8, pp. 7-29.

Apgar M. (1998) The Alternative Workplace: Changing Where and How People Work, *Harvard Business Review*, May-June.

Argyris C. (1998) The Emperor's New Clothes. *Harvard Business Review*, May-June.

Averill J. R. (1973) Personal Control over Aversive Stimuli and its Relationship to Stress. *Psychological Bulletin*, Vol. 80, pp. 286-303.

Baldry C. Bain P. & Taylor P. (1998) 'Bright Satanic Mills'; Intensification, control and team Taylorism. In: *Workplaces of the Future*. (Ed. by P. Thompson & C. Warhurst.) Macmillan Business, London.

Barker R. (1960) Ecology and Environment. In: *Environments: Notes and selections on objects spaces and behaviour*. (Ed. by Freidman & Juhasz) Brooks-Cole, Monterey, Ca.

Becker F. (1991) *The Total Workplace*. Van Nostrand Reinhold, New York.

Donald I. (1994) Management and Change in Office Environments. *Journal of Environmental Psychology*, Vol. 14, pp. 21-30.

Dunphy D. & Bryant B. (1996) Teams: Panaceas or prescriptions for improved performance. *Human Relations*, Vol. 49, pp. 677-699.

Evans G. W. and McCoy J. M. (1998) When Buildings Don't Work: the Role of Architecture in Human Health. *Journal of Environmental Psychology*, Vol. 18, pp. 21-30

Gibson, J. J. (1979) *An Ecological Approach to Visual Perception*. Houghton Mifflin, Boston.

Glass D. C. & Singer J. E. (1972) *Urban Stress*. Academic Press, New York.

Grimshaw R. W. (1999) FM: The Wider Implications of Managing Change. *Facilities*, Vol. 17 No.1/2, pp. 24 -30.

Grimshaw R. W. (1995) *The Impact of Design on the Use and Management of Housing for Young People*. UCE Press, Birmingham.

Goffman E. (1961) *Asylums*. Anchor Books, New York .

Kleeman, W. B. (1986) The Office of the Future. In: *Behavioural Issues in Office Design*. (Ed. by J. Wineman.) Van Nostrand Reinhold, New York.

Kay J. (1999) Quotes from the *Royal Society of Arts* Lecture on Ethics and the Role of Business in Society, London.

Kupritz V.W. (1998) Privacy in the Workplace: The impact of building design. *Journal of Environmental Psychology*, Vol. 18, pp. 341-358.

Lang J. (1987) *Creating Architectural Theory:The Role of Environmental Sciences in Environmental Design*. Van Nostrand Reinhold, New York.

Moore G. and Mitchell McCoy J. (1998) When Buildings Don't Work: the Role of Architecture in Human Health. *Journal of Environmental Psychology*, Vol. 18, pp. 85-94.

Murray P. (1999) How smarter companies get results from KM. In Mastering Information Management: Part Six, Knowledge Management, *Financial Times Supplement*, London.

Osterman P. (1994) How common is workplace transformation and who adopts it? *Industrial and Labour Relations Review*, Vol. 47, pp. 173-188.

Parker S. K., Jackson P. R., Sprigg C. A. & Whybrow A.C. (1998) *Organisational interventions to reduce the impact of poor work design*. HSE Contract Research Report, 196/1998.

Pedersen D. M. (1997) Psychological Functions of Privacy, *Journal of Environmental Psychology*, Vol. 17, pp. 147-156.

Phillips D. (1996) The Practical Failure of Architectural Psychology. *Journal of Environmental Psychology*, Vol. 16, pp. 277-284.

Preiser W. F. E., Vischer J. C. and White E. T. (1991) *Design Intervention: Toward a More Humane Architecture*, Van Nostrand Reinhold, New York.

Preiser W. F. E. Rabinowitz H. Z. & White E. T. (1988) *Post Occupancy Evaluation*. Van Nostrand Reinhold, New York.

Quinn R. E. (1988) *Beyond Rational Management*, Jossey-Bass, San Francisco.

Renn O., Webler T & Weidemann P. (1995) *Fairness and Competence in Citizen Participation: Evaluating models for environmental discourse*. Kluwer Academic Publishers, London.

Seligman M. E. P. (1974) Depression and Learned Helplessness. In: *The Psychology of Depression: Contemporary Theory and Research*. (Ed. by R. J. Friedman & M.M. Katz) Winston-Wiley, Washington, DC. pp. 83-113.

Steele F. (1973) *Physical Settings and Organisational Development*. Addison-Wesley, Reading, MA.

Sundstrom E. (1984) *The Psychology of the Workplace*. MIT Press, Cambridge, Ma.

Vischer J. C. (1989) *Environmental Quality in Offices*. Van Nostrand Reinhold, New York.

Westin A. (1970) *Privacy and Freedom*. Ballantine, New York.

Wineman J. (Ed.) (1986) *Behavioural Issues in Office Design*. Van Nostrand Reinhold, New York.

13 Managing the Journey to Work

Ashley Dabson, MCI Worldcom Ltd

The Government's White Paper on Transport, "A New Deal For Transport - Better For Everyone", published on the 20th July 1998, is an attempt by government to provide an integrated transport policy for the UK. This White Paper is not about 'banning the car' it is about tackling the problems of congestion on the roads and pollution. To do this a number of measures are put forward. One of the most important was the need for local authorities to produce Local Transport Plans and implement these through partnerships between local authorities, businesses, public transport operators and users. It has the aim of trying to 'persuade people to use their cars a little less and public transport a little more'.

Corporate Commuter Management

Corporate Commuter Management (CCM) may be defined as the conduct of affairs in a manner that ensures that the place of work remains accessible to staff, customers and visitors in a safe, convenient, economic, sustainable, and environmentally advantageous way. Green Commuter Plan's (GCP's) comprise three parts:

1 a survey of both existing commuter patterns and staff views
2 an analysis of the results of the commuter survey
3 a set of proposals designed to:-
    - reduce the need to commute and when necessary,
    - commute in a manner that is consistent with the aims identified above

Worton Grange Commuter Plan

The reason for preparing the Commuter Plan (April 1997) related to the compulsory acquisition of, what is now, part of the Compaq site car park for the construction of the A33 Relief Road between Junction 11 on the M4 and Reading Town Centre. In order to obtain planning permission for the replacement of the 270 car park spaces lost, Reading Borough Council as the Local Planning Authority insisted that a Green Commuter Plan be produced. The Compaq UK Management Team adopted a pro-active stance based on an assessment of the difficult situation prevailing at the time more than the requirement to produce such a plan. At this time the 1803 employees based at the site used 1191 on site car park spaces and 250 off site spaces. The off site spaces were inconvenient and expensive as a staff bus service was required. In addition, the traffic conditions in the South Reading Corridor between the Town Centre and Junction 11 on the M4 were deteriorating. As the newly developed offices in the Worton Grange area filled up during the period 1995 to 1998, peak hour delays of 30 and 45 minutes to exit the car park would regularly be encountered. Queues along the A33 and the M4 particularly at Junction 11 also lengthened. As a result the company's staff found these traffic conditions highly inconvenient and very stressful.

The Survey and Analysis

The survey was conducted by means of a questionnaire sent to the 1803 digital staff based at the Worton Grange facility. The survey was conducted during 16th and 17th December 1996. Of the staff surveyed 803 (44.5%) responded. Although 65% of the respondents to the Worton Grange survey indicated that they did not need their car during the day for work, 74% indicated that they did not have an alternative means of transport. What was interesting was that only 17% of those using their car to commute, but not needing their car for work, indicated that they were able to change their mode of transport. This reflects the perceived inconvenience of the alternatives to the car at the time.

The survey also dealt with a number of other issues including arrival and departure times, working hours, residential location, journey times, current and possible modes of commuter transport and the level of car dependency. The analysis of the results indicated a spread of arrival and departure times with some 40% of employees arriving before 8:30 a.m. and some 10% still in the building after 6:30 p.m. It also indicated that some 50% of employees lived within 30 minutes while 10% had a journey time of more than 1 hour and 15 minutes from Worton Grange.

The survey analysis also indicated that 30% of the respondents could change their work style from being office based to being tele-commuters comprising home and mobile working. Home workers may be defined as those who base their work at home and as a result have no dedicated office space. Experience suggests that within a corporate environment staff spending all of their time at home create management and team maintenance problems as gradually they see themselves as individual contributors and not part of the company. Therefore space and time are needed in the office for the purposes of control, direction and team building.

Mobile workers may be defined as those who at different times work from a number of different locations. These may be other offices, customer sites, hotels, business centres and the office from which they are supported. This category of staff will be in the office more than home workers but also not enough to have an assigned desk.

For these reasons both home and mobile workers have reduced office space needs. In terms of time this has been averaged at 10% for home and 33% for mobile workers. In both cases this space is shared. Depending on the tasks performed this means that unlike assigned space where significant amounts are left unoccupied while staff are out, shared space is sized to those in at any one time. Therefore, occupancy levels for shared space are significantly higher than for assigned space where the tasks being performed demand attendance at meetings or travel. In this situation buildings supporting home and mobile workers are used more intensively than those supporting staff who perform broadly similar tasks but have assigned desks. Indeed, it is the cost savings resulting from the intensification of the use of space by home and mobile working that have to a large measure been responsible for its adoption.

The Commuter Plan

The Commuter Plan was produced in April 1997. Firstly, it looked at the responses from the survey and then proportioned the results up to the total population that could be supported by the building. Using the analysis of the survey the Plan identified those who indicated that they could and would be permitted to change their work style. It then looked at the number (1699) of people in the building at the time of the survey. Subsequent electronic entry checks supported these numbers (Table 3).

Table 3 Total survey participants.

Staff status	Number	Number out	Number in
Home workers	13	12	1
Mobile workers	113	78	35
Staff with assigned desks	2,064	401	1,663
Total	2,190	491	1,699

The Plan indicated that there should be a limited increase in home (up to 104) and mobile (up to 182) working, allowing an extra 125 new staff to be employed. Therefore, although the number of desks in the building was planned to remain at 2100 only 1705 of them would be in use at any one time. This Plan also meant that the building would be able to support 2315 people (Table 4). Using the survey results the Plan produced a current and proposed modal split for the journeys to work (Table 5). The modal split was derived from questions in the survey aimed at identifying how many people could change their mode of transport for their journey to work. The car parking need was calculated as a residual number after the other modes had been planned in. Thus, 1154 parking spaces were provided. This was made up of 1089 for staff and 65 for visitors.

The commuter plan put forward 22 proposals, all but one of which has now been implemented. These included setting up a Worton Grange occupiers group, providing touch down space and support for mobile workers and visitors and retaining business centres at Hedge End (Solent) and Basingstoke. It also included providing cycle and motorcycle facilities (secure sheds, locker and shower rooms), services (dry cleaning etc.) and part sponsored bus services and a proposed park and ride south of junction 11. Finally, it also included a number of infrastructure improvements including improved on-site circulation and decking a car park to replace the spaces lost to the new A33 Relief Road.

Table 4   Number of people building able to support.

Staff status	Number	Number out	Number in
Home workers	104	94	10
Mobile workers	182	121	61
Staff with assigned desks	2,029	395	1,634
Total	2,315	610	1,705

Table 5   Proposed modal split.

Mode	Survey	Planned
Walk	33	42
Bicycle	49	85
Motorcycle	33	46
Bus	65	200
Coach	99	146
Car share	66	87
Taxi	6	10
Car	1,348	1,089
Total	1,699	1,705

Corporate Change and it's Impact on Commuting

By 1999 the desk capacity of the Worton Grange site had been increased to 2197 with 1995 of them either assigned to office based workers or occupied by home and mobile workers. This reflected the staff reductions that resulted from the integration of the companies. The modal split of the journey to work was surveyed (Table 6). The cycling and motorcycling numbers were down owing to 240 redundancies being made among the Engineering groups where the cycling and motor cycling culture was strong and the nature of their work meant that they did not need a car during the day. As other business groups have grown the total numbers supported by the building have only marginally changed to 2214. However, owing to people working elsewhere, those in the building at any one time have decreased to 1443.

Table 6   Actual modal split.

Mode	Survey	Planned	Actual
Walk	33	42	40
Bicycle	49	85	32
Motorcycle	33	46	28
Bus	65	200	150
Coach	99	146	112
Car share	66	87	86
Taxi	6	10	6
Car	1,348	1,089	989
Total	1,699	1,705	1,443

Coporate Human Resource Policies and Commuting

Changing business models, corporate outsourcing, acquisitions, growth, downsizing, densification, and tele-working all result in some form of portfolio change. The net effect has been to rationalise and reduce it. Buildings close and staff either uproot with all of the implications to spouse employment, schooling and other domestic arrangements or commute further. Many choose the latter.

During the 70s and 80s Digital grew rapidly with new locations being opened every year. During the 90s the business struggled and as a result costs had to be radically reduced. The result was 900 redundancies, and drastically increased commuting distances for the 420 retained personnel affected by building closures. This increased the number of employees on the site, causing the car park need to exceed the car park capacity. This situation was resolved by taking 250 off site car park spaces. Since the take over of Digital by Compaq in 1998 further changes have occurred due to the need to rationalise the 'back office' functions and consolidate the portfolio. In short, businesses change and these changes alter the demand, direction and modal split of commuting.

Commuting Information and Service

It is recognised that HR policies have a significant influence on commuting, including:

- Company car policy (fully expensed cars do not encourage staff to patronise public transport while payments in lieu do).
- Relocation policies (generous relocation packages encourage people to move which reduces commuting distances while 'inconvenience' and 'home to work' mileage allowances encourage long distance commuting).
- subsidising and advertising public transportation.
- canteen opening times (early morning breakfasts spread the a.m. peak).

- Car share schemes (providing convenient parking and sponsoring car share information systems).
- Car pool schemes to enable people who do not need to use their car for their journey to work to have access to a car during the day.
- Home and mobile working packages (encouraging flexible working by allowing staff to have the necessary equipment at home and or the necessary mobile equipment).
- Bus services provided on a 'no cost' to the employee basis.

The Future

On a number of occasions it has been found that the public transport available and potential car share lifts are not used owing to a lack of knowledge of what is available, its timetables and connections. The most significant problems encountered when wishing to use public transport relate to service levels. For example:

- Only parts of the desired journey are covered.
- Time tabling does not match the desired departure times.
- Bus routes do not match the desired line of the trip.
- Delays and inconvenience associated with interchanging between routes and modes are often made worse by poor co-ordination between providers.
- Public transport struggles to keep to its published timetables.

As part of the Green Commuter Plan bus timetables were published and software was developed which via the Internet allowed respondents to provide details of their journey so that others travelling from the same direction and wanting lifts could be put in touch. Having encouraged this by offering the most convenient car parking spaces exclusively to those who share their cars, some 86 people currently use the car share system.

At relatively low cost the CCTV security and traffic cameras can be linked to the Internet. At peak times the cameras are used to monitor the levels of traffic so that staff planning their journeys are able to see what is happening on their computer monitors before they leave their workstations. At Worton Grange the results have been encouraging, tending to even out the worst of the traffic peaks.

Compaq's experience provides a number of pointers for the future:

Intesification and Deintensification

- Using information technology to promote environmentally sensitive ways of working.
- Developing park and ride bus schemes.
- Providing a comprehensive network of cycle ways, facilities and services that are attractive to cyclists.
- Encouraging local authorities to develop partnerships with transport operators and major employers in order to support and co-ordinate private commuter management initiatives.

At a more fundamental level, Compaq could adopt a more aggressive plan to encourage:-

- home working to drastically reduce the need to commute by allowing employees to work via the internet from home
- mobile working to reduce the need to commute at peak times by being able to work from a number of different locations at the convenience of the employee
- higher density working enabled by flat screen technology to reduce the demand for more space
- distributed team working from local business or tele-centres to reduce commuter travel distances

With these arrangements staff would come into the office on an infrequent basis using touch down desk facilities. As such facilities are shared the intensity of use would increase. For this reason the more the workers with assigned desks alter their work style and become home and mobile workers the more the demand for commuter transport. Home and mobile workers are as likely to come in for meetings as they are to use touch down desks. They also tend to commute by car as they often need them for multiple trips during the day. For this reason public transport is inconvenient and relatively expensive, as season tickets are inappropriate. This in turn means that more parking space per unit area is likely to be needed not less.

Further, such workers tend to come in for meetings. This drives up the demand for meeting rooms and meeting areas. As the density of people in meetings is higher than that for office space this again drives up the number of trips generated per unit area of space. Meetings, particularly large ones have a major impact on the car parking needs of out of centre locations such as Worton Grange. Although video-conferencing and conference calling reduce the demand for face to face meetings they do not eradicate it. For this reason Compaq has located its main meeting and conference facilities in London and has restrained the size of meeting rooms in Worton Grange to 50.

The introduction of flat screen technology will reduce occupancy costs by allowing employers to increase densities by up to 20%, using smaller desks to meet the display screen regulations. Again higher densities mean higher needs for transport per unit of space. Therefore flat screen technology will mean that office buildings need to have improved transport links and infrastructure.

Partnerships

Increasing the number of people in a building by introducing more home working, flexible working and/or flat screen technology means that the commuter demand increases. For this reason a policy of restraining car parking in non-central locations that are poorly served by public transport is more likely to discourage employers from increasing the population density or utilisation of their buildings than it is to significantly shift the modal split away from the car. Therefore, it is likely that policies of this type will inadvertently encourage further development and discourage the very technology and way of working, that can do most to reduce congestion and it's associated environmental problems.

Distributed team-working, tele-centres and business centres can all reduce the length of the journey to work. They can help firstly, to retain key staff who would otherwise find relocation or the lengthened journey to work unacceptable and secondly, retain a geographical presence for customers. For these reasons Compaq has a number of these centres across the country.

Park and ride schemes are normally associated with radial trips between town centres and peripheral car parks. This concept can be developed to include allocated car parking and bus stops at employment or shopping centres along these routes during weekdays. During weekends all the parking can be turned over to shoppers with a direct non stop service to the shopping areas. There is no reason why the park and ride concept can not be extended to link park and ride car parks along motorways or other trunk roads. This would mean that commuters (and others) would gravitate from suburban and rural hinterlands, to a park and ride located at a motorway junction. They would then travel down the motorway by coach to the park and ride car park on the motorway adjacent to their destination. Using this as an interchange they would then change to a bus that takes them to their destination. As the motorway park and ride car parks become interchanges they could be provided with basic retail and other facilities.

Conclusions

Commuting stems from the current way that socio-economic activity is located, organised and transacted. Therefore, public sector action has to be seen in terms of regional or even national competitive advantage or disadvantage. For this reason, local authorities, public transport operators and major employers should be encouraged to form partnerships. Responsibility for the adoption of 'green' working practices should rest with employers, both public and private, aimed at reinforcing government/employer partnerships and encouraging employers to produce commuter plans and manage their commuting.

In essence, partnerships are based on two premises. The first is the employer is in the best position to understand how their business may be organised differently so that it may be transacted in an environmentally sensitive way. The second is that government, particularly local government, is in the best position to manage and co-ordinate commuting. Experience in the Reading and Wokingham area would appear to support this approach as both authorities have realised that, although not easy, working with employers as partners can achieve much. The nature of this process is also one of experimentation with all of its associated risks. Therefore, mistakes must be accepted, lessons learnt and new ideas tried as public authorities and employers work together to achieve the environment desired by our community.

There is little doubt that the car travelling from the garage at home along a clear road to a car park at work is a convenient form of physical commuting. It provides a door to door mode of transport available at all times with a capacity to carry personal effects. Unfortunately, the most valued attribute of convenience is diminishing as congestion worsens. Further, driving requires concentration therefore, it is a time waster, as no other activities can safely be performed during the journey by the driver. For these reasons it is important that green commuter proposals are aimed at being convenient, and reducing the time wasted.

Tele-commuting can achieve both aims. The overall demand to commute is reduced by home and mobile working while distributed team working from business centres will reduce commuting distances. The rapid growth (from the current small base) of e-commerce supports this as an increasingly viable alternative. This means that existing space will be used more intensively.

Flat screen technology will enable building density (occupancy levels) to be increased. High densities mean fewer buildings with more commuting to and from each individual building occupied. Although public transport becomes more attractive as densities rise it also means that measures aimed at reducing on-site car parking in locations that are not well served by public and other modes of transport are unwise. This case study indicates that park and ride schemes have also yet to come of age particularly those aimed at inter-urban travel. It is also too early to tell what roles cycling and car sharing will have in the future.

Lastly, it is clear from this work that sustainability depends upon the establishment of a change tolerant legal and fiscal framework. This framework must encourage innovation and the establishment of partnerships between public authorities, public transport operators and employers. These partnerships should be based upon shifting some of the responsibility for commuter management onto the employer and making local authorities accountable for achieving a free flowing transport system that is safe, convenient, economic, sustainable and environmentally acceptable.

What is needed is an environment where proposals can be continuously generated, applied, tested and developed or discarded. The one off green commuter plan may be a useful tool to control car usage but is no substitute for on-going corporate commuter management co-ordinated by local government. If we are to be successful in reducing road traffic congestion and pollution then the need is to persuade local government to seek to control a little less and manage a little more.

Part Three

The Property Trail

14 Introduction to the Property Trail

The new business imperatives as described in Part One, have triggering property market adjustments which, in turn, are leading to shifts in the relative importance of the strategies employed within property management generally. The changing business, social and functional requirements for buildings, as described in Part Two, are leading to reductions in the aggregate demand for built space per employee, per customer and per unit of production. The residential sector is the major exception here, together with some parts of the leisure industry. As a result greater emphasis is now being placed on the better management and utilisation of the existing building stock rather than the problems of new building procurement alone.

Much of the middle-aged property, inherited from the 60s and 70s, continues to constrain businesses and building users significantly. These constraints are imposed principally by the spatial form, servicing regimes and poor specification of this stock, most severely in the case of the older deep plan, fully conditioned office environments in the private sector, and poorly designed and cheaply specified buildings in the public sector. High levels of premature obsolescence has resulted, driven by functional and locational factors rather than by age, physical deterioration and depreciation alone. So during the 90s attention has been directed to the adaptation of redundant buildings to new uses, on the one hand, and to the disposal of buildings that were surplus to requirements, on the other hand.

Recently, procurement priorities seem to be changing in many corporate organisations, with greater emphasis on the procurement of buildings for leasing rather than for owner occupation. Market and professional attention has switched to partnership procurement strategies, particularly those under the PFI in the UK and the DBFL (Design - Build - Finance - Operate) approach in the USA. So over the last twenty years along the Property Trail, the emphasis has tended to move from those issues shown at the top of Figure 20 towards those at the bottom.

The construction industry continues to be predominantly 'supply-led', preoccupied with production rather than the use of the building product. It has working out of an oddly outdated position where business clients and building users have been viewed simply as 'property tenants' rather than property customers. In the UK the 25 year lease is the major millstone to business.

"The UK property market when compared to virtually any other property market in the world, is the most uniquely un-orientated towards business" (Oliver Jones, Chief Executive of Citex Group Ltd., 1999).

Figure 20 Changing Property Priorities.

FROM   - Traditional Property Procurement Strategies

- Strategies for Property Consolidation
- Building Use and Management Strategies
- Corporate Real Estate Management Strageties
- Strategies for Flexibility and Modification
- Strategies for Re-Use, Mixed-Use, Change of Use
- Strategies for Property Adaptation
- Strategies for Disposal and Selective Demolition

TO   - New Procurement Partnership Strategies

This position is now changing fast. The construction industry is facing up to a more complex and diverse set of client demands for a wider range and greater variety of types of building space, to support changing functions, mixed use requirements at distributed locations. Overall these changes in demand seem to be leading to a future where grater emphasis is given to the needs of business for highly flexible arrangements for building occupancy, use, operation and management, with increasing preference for renting under shorter and more flexible leases, backed up by the occasional use of serviced offices, temporary accommodation and other informal arrangements, to meet the exceptional and unpredictable periods of demand. Organisational clients now need to weigh up business advantage of flexible tenure with less or no investment and disposal risks, against the disadvantage of higher operating and servicing costs that flexible property arrangements entail. So key questions at the start of the Property Trail will include:

- How should public and private sector organisations reposition their property and facilities strategy to face the uncertainties of the future?
- How might changes in technology and changes in the nature and organisation of work and leisure, affect expectations and demands for property towards 2020?
- How might future facilities provide better operational support to meet the 'real' requirements of the corporate and individual property consumer, and the needs of business and society generally?
- What design innovations and new forms of facility are needed to provide strategic support to the dynamic business and social needs of this new century?
- How at the urban scale might facility and infrastructure innovations provide more sustainable and supportive environment for business and social organisations of all kinds?
- How will all of these questions impact on the ways that property and facility performance is measured and managed in the future?

As mentioned above, there seems to be a widely shared consensus that property and facilities will need to become more flexible, adaptable and sustainable in the future. While resource flexibility is increasingly vital to meet the ever changing needs of organisations, property is by its nature, a highly inflexible and illiquid resource. Chapter 15 explores this dilemma and suggests that property portfolios need to become not only more physically flexible but also more functionally and financially flexible. It argues for the adoption of mixed portfolio positions in which core property, suitably designed to be more manageable and functionally flexible, should be supported by peripheral property, which is tenurally and financially flexible. Similarities between 'core and peripheral' property resource management and 'core and peripheral' human resource management are explored. It is argued that each property portfolio should be assessed as part of a resource audit to determine an organisations freedom of manoeuvre along physical, functional, financial and tenural dimensions of flexibility. A framework for evaluating the relative flexibilities of a property portfolio is proposed as a basis for assessing when, where and what type of flexibility is required.

It is only recently, and perhaps mainly as a consequence of PFI projects, that facilities managers have begun to make an input to the process of property design. Chapter 16 discusses the potential contribution of facility management knowledge and experience to the design process, particularly during the briefing stage. Unlike the views expressed in Chapter 12, it is suggested that user participation during design is of limited value, since users are likely to have little if any experience of the overall product or process of design. In contrast, the operational experience of facility managers could place them in an uniquely powerful position to bridge the gap between the knowledge of the hardware (real buildings, their services, maintenance and physical modification) and an understanding of the software (building utilisation and management). The ways in which the operational and strategic experience of facility management, both 'hard' and 'soft', might be 'designed-in' to new buildings and major refurbishments is a problem that is still to be tackled, not least the basis and form of the facility management brief itself. So an expert model 'FM briefing framework' for structuring the contribution of facilities management knowledge and experience, needs to be developed urgently.

In the past, the attention of architects, space planners and facilities managers has tended to focus on the dedicated workplace. Chapter 17 compares three alternative generic approaches to the implementation of new workplace practices. It describes how workplace design can be used as a tool to support an organisation's business strategy, acting as a catalyst to facilitate change, to improve work performance and to enhance end-user satisfaction. Chapter 18 explores the linkage between property performance and business productivity. It examines three key workplace metrics; cost, utility and quality, and their possible impact on productivity. The case for a shift in emphasis is made, away from a narrow view of productivity related to simple bench marking of premise's costs and utilisation, towards measures of productivity that link directly to business performance, efficiency and output.

The following chapter describes the changing priorities and practices of property management, both in Europe and North America. Chapter 19 gives a detailed account of shifts in the policies, functions and activities of corporate real estate management, with an assessment of the knowledge and skills that will be crucial in the future. The importance of both facility management and environmental management, within corporate real estate management generally, appears to have declined during the last few years. Perhaps FM's honeymoon period along the Property Trail, is coming to an end, suggesting the need for new FM performance metrics beyond 'cost cutting' and 'downsizing'.

> "In order to maximise the strategic contribution of Corporate Real Estate (CRE), emphasis must move to the metrics used by the core business competencies. In this way CRE managers can understand the objectives of the organisation and the organisation can appreciate the contribution that can be made by CRE".
> (Ron Adam, Chairman IDRC UK, 1999)

As discussed in Part Two, a significant shift in the demand for buildings is underway largely due to new working practices but also as a result of an increasingly demanding environmental agenda for businesses and property development alike. These changes are both reducing the extent and altering the nature of the demand for space. Chapter 20 looks at the potential for adapting existing building stock to meet the changing nature and structure of demand. It explores how existing buildings may be adapted to support new needs and how new buildings may be designed to support a variety of uses and functions. This should help to ensure more sustainable buildings, but the possible implications of these fundamental changes for the future form of property and facilities are only beginning to be grasped. The chapter concludes that a degree of physical redundancy, use ambiguity and flexibility could lead to a more adaptable and sustainable future for infrastructure, buildings and facilities. However, this will require radical change to the traditional 'land-use' planning systems that have been in operation over the last fifty years. The introduction of a more responsive, intelligent and permissive regulatory system of planning controls will be required, with which to manage the dynamic and complex patterns of future demand and use of property.

References

Adams R. (1999) The Corporate Real Estate 2000, Synopsis in Conference Proceedings: *Futures in Property and Facility Management*, University College London, London, p. 89.

Construction Industry Council (CIC) (1999) *Constructors Key Guide to PFI*, Thomas Telford, London.

Jones O. (1999) New Strategic Directions for FM, Conference Debate Transcript, *Futures in Property and Facility Management*, University College London, London.

15

Property Portfolio Dynamics: the flexible management of inflexible assets

Virginia Gibson, University of Reading

As the pace of change increases, organisations are seeking ways to become evermore responsive to the external environment. Technological innovation, global competition and the restructuring of political economies are but a few of the factors leading to a more unpredictable competitive environment. Numerous management initiatives have attempted to respond to this new environment. Downsizing, delayering, outsourcing, business process re-engineering (BPR), and core business focus, have all been used as ways to gain greater adaptability.

There is potentially a direct impact of these changes on corporate property portfolios. Downsizing, delayering and outsourcing have all lead to a reduction in the demand for space. BPR and the focus on core business, have led to the need for different types of space in different locations (Gibson and Lizieri, 1998). This has presented real challenges to corporate real estate managers who have been attempting to realign their existing portfolios to meet the new organisational requirements.

However, during this realignment phase it has become increasingly clear to corporate real estate managers that there is a need to develop workplace solutions which can continue to adapt to the changing requirements of the organisations. Flexibility is therefore one of the key issues debated by many corporate property and facilities managers: how can they gain greater flexibility from a resource that by its very nature is inflexible?

In searching for an answer to this question it becomes clear that the issue is multifaceted. For instance, what type of flexibility is being sought, physical, functional or financial? Is flexibility required for the whole portfolio or for only part? This chapter proposes a framework for responding to these questions, building on research undertaken during the last two years which has investigated the drivers to decisions made by managers of corporate property. The aim is to better articulate the dynamics of corporate property portfolios and demonstrate that as organisations evolve, so do their property resource requirements. Some requirements are long term by their very nature, possibly a headquarters facility, while others are increasingly transient. The level of involvement and management responsibility the organisation wants to take on will be very different depending where on the continuum the property sits. This can be incorporated within the debate in the management literature concerning core and non-core activities. Similar work undertaken in relation to the management of other resources, in particular the way human resources are now being grouped within major corporate organisations as core and peripheral workforce, also needs to be considered. One of the questions being addressed is whether these models can be applied to the property resource.

This chapter is divided into two main sections. First the debate regarding the flexibility of property assets is examined. This draws together ideas which have often not been considered explicitly in relation to flexibility. It shows how the three key property perspectives - physical, functional and financial - can all contribute to the overall flexibility of an asset from a user's perspective. The second section reviews two frameworks: one for evaluating the flexibility of an individual asset and one for examining the flexibility of the portfolio as a whole. These are developed as tools for corporate property managers to use in a strategic context in order to assess the strategic capability of the property assets within an organisation's portfolio. The key findings are then drawn together in a conclusion which highlights the lessons for corporate property managers.

The Flexibility Debate

The literature related to the flexibility of property assets is diverse and often disjointed. Given that property can, and probably should be viewed from a number of different perspectives this is not surprising. The field comprises professionals from very different backgrounds including property finance, development, construction and architecture. Each of these gives rise to a particular orientation. The differing perspectives have been articulated by a number of writers. Bon et al. (1987) developed a framework which grouped corporate property management activities into three categories: financial, physical and organisational use. Gibson (1995) argued that these same perspectives created difficulties for corporate property managers to take a strategic overview because of the inherent conflicts between the perspectives. More recently Horgen et al. (1999) describe the workplace in terms of four interrelated elements: spatial, organisational, financial and technological. They argue that the flexibility required within the workplace relates to the space, the infrastructure, the workforce and the work processes. What emerges from these writings is that examining an issue like flexibility within the corporate property portfolio is multidimensional and therefore needs to be considered from the various perspectives. From the point of view of a corporate property manager, the physical, functional and financial aspects of a property are those over which they have control and therefore any framework will need to consider these as sources of flexibility.

Therefore taking first the physical perspective, flexibility has been articulated in terms of building design including usable areas, modular floor plates, and the ability to change the internal configuration of space (Harris, 1996). It is contended that organisations need property assets which can physically adapt to changing business requirements (Harris Research Centre, 1994). There has been pressure on property developers and suppliers to build property products which are physically adaptable with the British Council of Offices (BCO, 1994) developing recommended specifications. The emphasis within this debate is on design, construction and services rather than the way in which the building is ultimately fitted out.

The functional flexibility debate includes much of the work on the alternative workplace. Duffy (1990) has argued that buildings are not designed to support the functions that are now undertaken in the contemporary office environment, but that designers are trapped in a view of office work at the turn of the century. Team space, meeting areas, free address areas, enclosed offices are all potentially needed within the modern office environment (Becker and Steel, 1995).

Another aspect of functional flexibility is the need to have spaces designed and used so that individuals and teams can relocate within the organisation with minimum downtime and cost. The cost of 'churn' is often considered to be an unnecessary expense if the properties were better designed and managed. There is evidence of organisations who have reduced the cost of churn by either introducing new working practices like hot-desking or more simply by standardising the office layout and equipment so that people not furniture move. (Strohm, 1995)

Financial flexibility has been examined in less detail. In the UK where the standard institutional lease was the norm until recently. Corporate real estate managers have argued that financial flexibility was only possible through freehold ownership (Avis and Gibson, 1995). It was maintained that only by owning the freehold did a corporate occupier have total control over what could happen to the property such as sell, sub-let, or mothball. A lease always had both contractual and financial constraints. More recently, major corporate occupiers appear to have found financial flexibility through the emergence of the serviced office sector (Gibson and Lizieri, 1999). Although the cost of occupation may be high, the financial flexibility, that is the ability to use the space on a very short-term basis or exit quickly, is seen as a priority for some activities and therefore worth the expense.

The problem with much of the literature is the lack of differentiation between flexibility within a single building and flexibility across a property portfolio as a whole. What has not been debated is how a corporate property manager might assess the flexibility within the portfolio as a whole and to identify in which properties flexibility is required and in which properties it may not be required.

Modelling Flexibility

In order to understand the flexibility within a portfolio, it is first necessary to develop a framework for examining the flexibility provided by a particular building and therefore making it possible to assess the degree of flexibility that is provided as a single asset. If one is taking a strategic approach to the management of property, assessing flexibility is an essential component in understanding strategic capability (Johnson and Scholes, 1997). Drawing on the ideas presented above, Table 7 suggests a few of the criteria which might be used to evaluate an asset. This analysis would be required to understand what degree and type of flexibility is available within each of the buildings within a portfolio. In many ways this is a resource auditing perspective; assessing what the capabilities of the existing portfolio actually are without any real judgement as to whether the flexibility is appropriate or necessary.

Therefore corporate property managers have to consider a more strategic approach. When is flexibility necessary and when is it non-essential? It is clear that some of the flexibility discussed above is only achievable at extra cost. For instance, establishing standardised office layouts throughout a large office building often means substantial investment in new furniture and equipment. Using serviced office accommodation has a much higher cost over a long-term period than renting a property on a standard lease (CIPS, 1998). If the flexibility within the building is not required for the current or future organisational activities then the extra investment or cost will be unnecessary.

Table 7   Criteria for Asset Evaluation.

Dimension	Issue	Criteria
Physical	Ease of reconfiguring space Efficiency of space Physical impediments to change	Cost (and time) of change % Net to Gross Area Ceiling heights; Bay widths
Functional	Variety of activities/processes undertaken Ability to support movement of staff Ability to change use	Balance of uses within building Standardised furniture/equipment Building Class (i.e. B1)
Financial	Ability to vacate Ability to let/sub-let	Contract length and exit costs Potential for income

This is where drawing on the other resource literature is appropriate. In human resource management, the focus on core business and the delayering of organisations lead to an examination of the employment contracts with the workforce. The concept of a core and periphery workforce to underpin the business has been developed to try and articulate where labour force flexibility was needed (IPM, 1996). The core workforce would be highly skilled staff and employed on long term contracts. They would expect to be broadly based in order to respond to a variety of tasks and problems within the organisations. In other words, they would be expected to be functionally flexibly and they would be rewarded for this through high pay and additional perks. This core workforce would then be supported by layers of periphery labour. These include individuals on part-time and short-term contracts, contract and out-sourced labour and agency workers all of whom could be brought in and out of the organisation in response to the stage in the economic cycle and the need for particular skills for short term periods.

This concept has been applied to an organisation's corporate property portfolio in order to gain a greater understanding how overall flexibility might be managed across a portfolio. Gibson and Lizieri (forthcoming) developed a three-tiered approach to examining a corporate property portfolio in terms of what might be seen as core and periphery property requirement. This is shown in Figure 21.

Figure 21 Three-tier Approach to Examining a Property Portfolio.

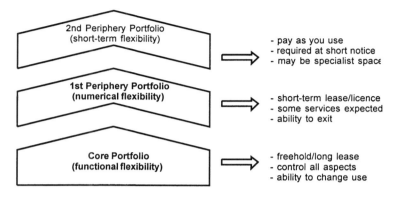

At the centre of an organisation's requirements would be the core portfolio. This is not intended to be mapped on to core activities or core staff, but to reflect an assessment by the corporate real estate manager of buildings within the portfolio, which are considered to be needed by the organisations for the long term. This could include facilities which are considered strategically located (manufacturing facilities), landmark buildings which embody the history and culture of the organisation (headquarters buildings) and space relating specifically to the organisations source of competitive advantage (research and development facilities). These are the buildings the organisation would be willing to invest in, but also with a high degree of control to adapt them as and when the organisation changes. In ownership terms they are likely to want to own these facilities on a freehold or long lease hold basis. The key flexibility issue would relate to functional flexibility, the ability to change the use of the building, and by implication some physical flexibility. Financial flexibility in terms of the ability to exit quickly is less important.

The first level of periphery property is that where numerical flexibility is required. As the demand for products or services fluctuates over the business cycle, the organisation will want to be able to service that demand in times of boom but to reduce the costs in times of recession. The key issue here is having some functional flexibility in order to allow for marginal growth within the building but more importantly the ability to exit the financial contract at particular points in time. The emergence of the shorter lease in the UK with break clauses would appear to be the way in which properties in the first level of periphery portfolio might be held.

The second level of periphery portfolio relates to the requirement organisations have for very short-term space. Speed of entry and exit are paramount and therefore financial flexibility is the most important. There are two types of space which appear to fall into this category. First, specialist spaces like training facilities which are used infrequently throughout the year by the organisation. This demand is met through a range of properties like hotels, conference centres and universities. Secondly, there is a growing need for generic office space to house overflow activities on a short-term basis or entry into new markets prior to establishing a more permanent presents. This demand is being met by the serviced office providers who are able to facilitate total flexibility both in terms of space and services.

But there is another dimension to the flexibility debate. When an organisation occupies property there are numerous servicing requirements which need to be put into place, such as design maintenance and facilities management. One of the issues as we move into the periphery layers of the portfolio is the degree that physical, functional and financial flexibility may be impeded by the need to also set up and establish a these management service. Therefore the two periphery layers within the portfolio imply that as well as space, services are also being provided by the 'property provider' or are being outsourced to another partner. This would relate closely to the use of contract workers who would not get company benefits such as company pensions or training, but rather would be expected to provide these for themselves.

Although in practice there is no clear divide between the three levels, the model does provide a framework for examining the organisation's property portfolio in a strategic way. The balance between the layers will be different for every company depending on the sector and competitive situation but in recent work the estimate has been 60% core supported by 40% peripheral (Gibson & Lizieri, forthcoming). The model therefore helps corporate property managers to focus their attention on the part of the portfolio which is perceived to add the greatest value. Time, energy and resources can be invested in the core portfolio while working with other types of space and service suppliers who can provide a more appropriate solution for the periphery portfolio.

Conclusions

Flexibility in property comes from a variety of sources. The construction and design of the basic structure establish the degree to which it can physically adapt. The way a building is fitted out internally, the balances of space uses and the degree of standardisation of equipment will influence its functional flexibility. Finally, the type of legal contract on which the property is held and the level of services offered by the property provider would influence its financial flexibility. A corporate property manager should be able to assess each of the assets along these dimensions.

However to really understand if a resource meets the current and future needs of an organisation, an analysis should be undertaken examining the type of flexibility required in different situations. As with human resources, different types of flexibility are required by the organisation at different times and to support different activities and functions. At the heart of an organisation's property requirements may be its core portfolio which is functionally adaptable which is supported by periphery property which provides, in particular, financial flexibility.

Property and facilities manages have often not seen their role as being similar to other resource managers. This paper has demonstrated how both the ideas in strategic management and in other resource management areas can be adapted to allow property and facilities managers to take a more creative approach to the management of what are perceived to be inherently inflexible assets.

References

Becker F. and Steele F. (1995) *Workplace by Design*, Jossey-Bass Ltd. San Francisco, CA.

Bon R., Joroff M., & Veale P. (1987) *Real Property Portfolio Management*, The Laboratory of Architecture and Planning, Massachusetts Institute of Technology, Massachusetts, US.

Chartered Institute of Purchasing and Supply (1998) *Breaking Down the Barriers to Entry*, CIPS, UK.

Duffy F. (1990) *The Responsive Office: People and Change*, Streatley-on-Thames: Steelcase/Polymath.

Frost Y. (Ed) (1993) *The Flexible Workplace, Future Work Forum*, Henley Management College, Henley.

Gerald Eve (1997) *Overcrowded, Under utilised or Just Right?* Gerald Eve/RICS, London.

Gibson V. & Lizieri C. (2000) New Business Practices and the Corporate Real Estate Portfolio How responsive is the UK property market? *Journal of Property Research*, forthcoming.

Gibson V. & Lizieri C. (1998) *Business Reorganisation and Working Practices - Trends and Real Estate Implications*, Research Paper, Right Space Right Price, London, RICS Books.

Gibson V. A. (1995) Is Property On The Strategic Agenda? *Property Review*, Vol. 5, No. 4, pp. 104 -109.

Harris Research Centre (1994) *Tomorrow's Workplace*, Richard Ellis, London.

Harris R. (1996) *Less a Castle more a Condominium: Taking a look at the office of the future*, Gerald Eve Research, London.

Horgen, T. H., Joroff, M. L., Porter, W. L. &Schon, D. A, (1999) *Excellence by Design: Transforming workplace and work practice*, Wiley, New York.

Institute of Personnel Management (1986) *Flexible patterns of work*, (Ed. by Curson, C.) I.P.M., London.

Johnson G & Scholes K. (1997) *Exploring Corporate Strategy: Texts and Cases*, 4th edn. Prentice Hall, London.

Lacity M. C., Willcocks L. P. & Feeny D. F. (1995) IT Outsourcing: Maximize Flexibility and Control, *Harvard Business Review*, May-June pp. 84-93.

O'Mara M. A. (1997) Corporate Real Estate Strategy: Uncertainty, Values and Decision Making, *Journal of Applied Real Property Analysis*, Vol. 1, No. 1, pp. 15-28.

Strohm P. (1995) The quest for efficient space, Office Trends 1995, Supplement to the *Estates Gazette*, 3 June, Vol. 9522, pp. 10-12.

16

The Impact of FM on Briefing and Design

Stephen Brown, The Charter Partnership

Customer satisfaction has remained a major focus of many industries throughout the last decade. The reconciliation of expectation and realisation, whether related to a service or a product, determines the level of satisfaction achieved. The construction industry is no exception and this has been recognised by numerous Government and industry initiatives.

Expectation is a complex and composite package of objective and subjective criteria. Some are expressed, but equally many are simply preconceptions. Realisation is the reality; it is frequently only evident a long time after the design process in the conceptual sense is over and often when the physical product is approaching completion.

The design process is itself, too often a continuum starting with major strategic conceptual decisions and culminating during the build process with operational and technical design, usually with little or no feedback. Much work during the late 80s and early 90s was directed towards the briefing process and the subsequent techniques for managing 'design' (Nutt, 1988; Preiser, 1993; Atkin et al. 1996; RIBA, 1997). More recently, the emphasis has been directed at construction and the operational or technical level of design. Recent studies have shown that architects for example, still have little involvement with management issues or strategic decisions (Bicknell, 1998). It is also clear that many projects exhibit minimal input into the critical strategic brief and, with undue haste, efforts are concentrated on the physical brief.

The 1999 Construction Industry Board survey indicates that satisfaction levels with both consultants and contractors has improved over a five year period. It does not however quantify the remaining absolute levels of dissatisfaction in relation to the design process.

Research involving a survey of building employers (Brown, 1998) showed that typically 10% of all projects are deemed to have failed in some respect before construction begins. It does not suggest that employers are knowingly allowing defective design to commence on site, but that simply the reality and consequences of a design is not apparent at that point of the process. Causation of this high level of failure is multi-fold, but the primary factors can be summarised:

- Needs change during a project – an euphemism for an inadequate and inflexible brief.
- Failure to fulfil the functional needs of users and subsequent managers.
- Missed time and cost targets.

The potential for projects to encounter difficulties during construction is clearly high, but there is little excuse for design stage failure. It is disturbing that a significant number of employers report failure rates much higher than the median 10% (up to 60%) in the three key goals of Function, Finance and Timescale. Figure 22, illustrates that although a significant number of employers report low levels of dissatisfaction, there is an equally significant body experiencing consistently high failures levels.

Figure 22   Failure Levels from 1998 Survey of Building Employers (Brown, 1998).

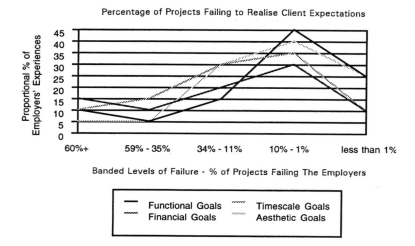

There have been a number of responses to these issues. User participation in the 1980s was heralded as a methodology for ensuring user satisfaction and adherence to operational functional demands. But 'Participation' has often resulted in expensive briefing processes and it has been demonstrated that involvement does not necessarily equate to satisfaction. Survey work has indicated that high participation levels in many organisations, particularly public sector, is paralleled by equal levels of dissatisfaction on completion of the project (Becker, 1991).

The response of the UK Government has been the promotion of the Latham and Egan reports and these have clearly identified goals for the construction industry as a whole. However they contribute little to the reconciliation of expectation and realisation in terms of the design process. The bringing together of the design team with the contracting team and with the employer's team is commendable, but clearly from the examples of participation outlined above, this is not a certain remedy, as continued failure levels demonstrate.

The UK's Ministry of Defence's Construction Supply Network Project (CSNP) is a direct pilot procurement technique derived from the Egan ideals. Utilising the concepts of open communication and design group clusters, it is endeavouring to deliver the targets set out by Egan. It involves a wide range of academic and practise based organisations to advise, monitor and record the effectiveness of the two pilot projects. A recent report suggests that the Prime Contractors however are struggling with the cultural issues involved (Tavistock Institute, 1999). The CSNP design processes engage two highly commendable ideals:

- Early access by the design team to the client/user.
- A project brief developed from the client's strategic goals into a form promoting understanding and a reflection of user requirements.

The toolbox of supply chain management techniques developed, nevertheless, retains an intensity of effort in the physical delivery. Most importantly the process still retains the traditional interpretation of the client brief and the space budget, by traditional singular professions. These include architects, surveyors and engineers, all of whom historically have a focused and restricted education and agenda. While it does create the potential for harmony, shared goals, values and motive for all involved in the project, it fails to specifically identify one key participant – the facility manager. Research shows that facility management is rarely involved as an integral part of the design process, with the possible exception of PFI. Consequently the high level of building failure for subsequent managers identified in recent studies, is not unexpected.

Figure 23 The Gap to be Bridged.

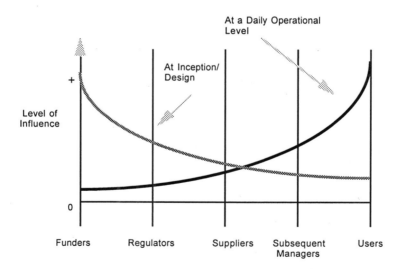

There remains a canyon between the initial instigators and the ultimate users of property. Attempts to bring together users in participatory ways is of restricted value. Frequently the interface between employer and the design team falls between an employer's representative and design team leader. The former frequently is inexpert and struggling with their core function. The latter often lacks any in depth knowledge of the client organisation and is struggling with time and cost agendas for their own organisation and the project.

As long ago as 1934, it was pointed out that poor building layout added to production cost, 15% in the case of the manufacture of cotton goods for example (Bossom and Cain, 1999) and yet the relationship between building design and function so often remains tenuous. Such problems may easily be dismissed as issues of communication, but although this may be true, the consequences remain un-resolved. Figure 23 indicates the hypothetical gap between those with influence and interest at inception and those at project completion. Similarly the project cycle sees shifts in the positioning of all project stakeholders in terms of 'power versus interest'. As long as this gap is maintained the resolution of true communication issues will remain elusive.

Facility management can potentially be both a bridge builder and translator. The canyon between instigators and ultimate users can be filled by FM having an informed understanding of the organisational process but without the preconceptions and hidden agendas frequently carried by employers, design teams and contractors alike. The work of Whorf and Sapir has shown how language and indeed culture is a barrier to understanding (Davenport, 1994). Levels of knowledge further compound the problem. Two-way translation of all involved in the design process and common language is key for there to be a genuine understanding of expectation.

Figure 24 The Interfaces of Facility Management.

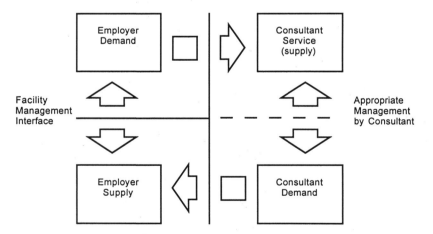

Increasingly professionals and employers are being encouraged to concentrate on core business and to specialise. This can only exacerbate the communication gulf. The management of the internal and external interfaces described by Figure 24, are essential and increasingly critical. The reconciliation of supply and demand within the employer organisation must be managed if internal expectation is to be fulfilled. Similarly the demands of the consultant must be managed in relation to the supply potential of the employer, in terms of information flow, to ensure an appropriate briefing process.

Finally if the employer's demand is to be satisfied then supply by the consultant, in terms of media, knowledge and methodology, require equal control.

Briefing has been acknowledged as a fundamental element of the design process but the understandable holistic brief remains elusive. Emphasis has been placed upon physical solutions and cost. The introduction of quality systems, together with value engineering still fails to address the root causes of dissatisfaction. Similarly procurement innovation is worth little if teams remain a network of isolated professions. It has been said (Duffy, 1995) that design is the process of matching hardware (physical and structural) to software (people and process). Too often employers and design teams move with excessive speed from strategic brief to solution, giving rise to irreversible positions and inflexibility.

Facility management has the opportunity to bridge these gaps for the benefit of all involved. Further development is required in two key areas:

- An accessible knowledge base.
- A clear and structured model for the facility brief.

The car industry uses 'non-sequential engineering' where all designers have access to the same data. The construction industry has each team member independently resourced and frequently working in isolation. Facility management has the potential to develop a broad knowledge base, spanning both hardware and software. Tacit knowledge abundant upon the completion of a project can supplement later Post Occupancy Evaluation (POE) data to provide a significant resource to input into a design process.

Employers and professionals both admit that experience remains the primary basis for briefing. Both cite recorded information as the 'last port of call' (Brown, 1998). The encoded nature of much data and poorly managed accessibility are primary reasons for the low usage. More than a technical 'fix' is needed however, and a culture change is essential.

The integration of facility management within the project team as a strategic resource, not simply a 'bolt on fixer', is fundamental to any serious attempt to close the gap between expectation and realisation. Figure 25 positions the facility manager in terms of current influence and role against a project time line. It also shows the potential to expand influence beyond the operational, and develop involvement earlier in the time line.

The growth of risk assessment and value engineering has been QS and construction led. There is, however, little evidence of 'use engineering', despite the protestations of many employers that they are indeed seeking holistic improvement. Many clients have begun to appreciate the potential that strategic facility management can provide within their organisation, but few have acted upon this belief.

Acknowledged failures in the areas of cost and time have encouraged management responses in the form of clear identifiable project control systems. Attempts to improve functional and qualitative issues have generally resulted in more communication but unfortunately without the anticipated improvement in understanding. The FM brief remains under developed.

The primary reason would appear to be the lack of any acknowledged model for this brief. The role of facility management has developed rapidly to the extent that it has become 'all things to all men'. Employer perception ranges from 'janitorial services' to strategic property asset management.

Figure 25 Future Directions for Facility Management.

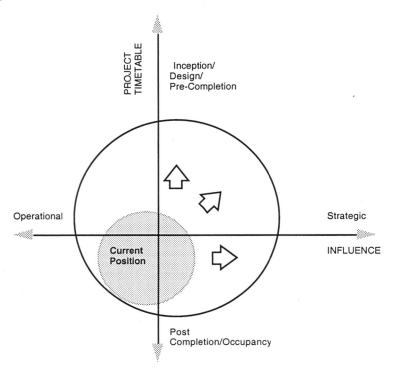

Traditional singular professions, such as architects and surveyors, have little knowledge of FM. The majority believe that FM issues have little impact upon their clients, their service or the design project as a whole. Responses are, however, polarising. Although many singular consultants are concerned that another member within a design team will further marginalise their personal role, there are those that are embracing FM. Cynically, it could be argued that these consultants are 'going with the flow', but most reasonably respond that the FM approach provides for the genuine resolution of many long standing issues. Facility management is, within the design process, as much an approach as it is a discrete profession. Indeed it has been referred to as a profession of professionals.

Development of an acknowledged model facility brief has the potential to clarify the role, and demonstrate the benefits that FM can bring to the design process and all involved with it. A contributor's power and influence at the early stage of a project is not always appropriate to the level of interest or potential contribution, and this is particularly evident in the FM arena. It is presumed that designers sincerely wish to close the gap between the expectation of all stakeholders and the reality of the completed project. Similarly it is presumed that employers wish to satisfy the total business demands of their organisation, and not just one vocal section. The current range of motives and agendas within all contributors, however, makes these presumptions somewhat naïve. Without strategic linkage and management of all aspects of supply and demand throughout the project team and its stakeholders, progress is unlikely. The plethora of Egan spawned 'Best Practice' initiatives, aimed mainly at procurement, will become mere window dressing if fundamental issues of briefing are not addressed.

It is clear that no one profession has the ability to reconcile all of the conflicting demands that inevitably arise during the life of the project. But the inclusion of FM as an equal member of the design team can act as a resin to gel together the matrix of professions involved. The payback for the team is clear. facility management has a potential to close the gap between expectation and realisation.

References

Atkin, Clarke &Smith (1996) *CONSTRUCT IT - Benchmarking best practice briefing and design*. Construct It Centre Of Excellence University Of Salford.

Becker F. (1991) *The Total Workplace*, Van Nostrand Reinhold, New York.

Bicknell C. (1998) *Architectural Facilities Management*, MSc Report, University of Westminster

Bossom, Lord Alfred (1999) quoted by Cain, C. in Building Down Barriers, *Faculty of Building Journal*, Feb.

Brown S. (1998) *Communication Interfaces - A Gap between Expectation and Realisation in Building Projects*, MSc Report, UCL, London.

Davenport G. (1994) *Essential Psychology*, Collins Education.

Duffy F. (1995) *Facility Management*, Vol. 12, No. 3.

Nutt B. (1988) The Strategic Design of Buildings, *Long Range Planning*, Vol. 21, No. 4, pp. 130-140.

Preiser W. (1993) *Professional Practice In Facility Programming*, Van Nostrand Reinhold, USA.

RIBA (1993) *Strategic study of the profession phase II, clients and architects*, RIBA publications, UK.

Tavistock Institute (1999) Interim Report - summary article, - *Building magazine*, 19/03/99.

17 Strategic Workplace Design

Kirsten Arge, Norwegian Building Research Institute

Over the last ten years interest in and implementation of different forms of new and innovative office solutions, or as Franklin Becker calls it, alternative officing (AO), has increased dramatically. Alternative officing has by now become an important tool in real estate and FM space planning and design. Books and articles in Harvard Business Review and other well-known business magazines, have helped to legitimise alternative work-place solutions and to map their potential cost savings to corporate leaders.

Until a few years ago, traditional organisational management literature did not acknowledge the relationship between work place solutions and organisational change and development and productivity. Lately, however, several management authors have included this aspect in their books on managing change, supporting team effectiveness, etc. (Sundstrom, 1990 and Pfeffer, 1997). This clearly demonstrates the actuality of AO in strategic organisational thinking at the moment.

This chapter describes the findings from a study of six Norwegian organisations who have introduced alternative workplace solutions. Among them are some of Norway's largest organisations. On the strategic level, the key research questions were:

- Do senior managers in Norwegian knowledge-based organisations have strategic objectives with office change?
- Are alternative workplace solutions used as a strategic tool to support organisational change and development?
- Do senior managers understand how alternative workplace solutions can be used as a strategic measure to add value to the organisation?
- Or are alternative workplace solutions only a question about introducing more cost efficient workplace solutions, to build and to manage as a facility?

The research adopted an explorative comparative case approach. Data collection was based on structured personal interviews, secondary sources as documents and drawings, as well as visits to inspect the workplace solutions. The interviewees in the organisations were top managers on the strategic level in business, personnel and facilities management departments, as well as project managers and architects involved in the work place design projects.

Theoretical Framework and Definitions

Duffy has suggested three organisational drivers behind office development: Grade & status, Function & task and Process & performance (Duffy, 1997). Grade & status favour cellular offices and size by grade or status. Function and task driven work place solutions tend to favour standardisation in order to support adaptability, often called universal plan offices. Process & performance driven work place solutions are challenging both the use of time and space, and the ownership of space. Process & performance driven work place solutions in Duffy's terms include both team or group work offices, and non-territorial offices/hotelling.

Becker includes altogether six types of 'alternative' workplace solutions in his term Alternative Officing (Becker, 1999).

- Universal plan offices/workstations.
- Activity-setting environments.
- Team/collaborative environments.
- Non-territorial/unassigned offices/workstations.
- Home-based telecommuting.
- Virtual officing.

The key characteristics of universal plan settings are according to Becker, settings with the same size workstations or footprint, regardless of rank, hence the term universal plan. The intent in these approaches to space planning and allocation is that the individual becomes mobile inside the office, choosing to work in particular settings that are designed and equipped for a particular activity or task, hence the term activity settings. On the other hand, team or collaborative environments have primarily involved the introduction of furniture and equipment designed to support collaboration and communication among relatively small groups numbering 5-25 people, while also supporting individual work. The essence of most 'team' environments according to Becker, is tables, whiteboards, tack surfaces and storage units designed to be easily re-configured as team members come together and apart. Electronic infrastructure which enables sharing and display of information have also become a part of collaborative work environments.

Two of the organisations in the Norwegian study had chosen universal plan settings in their new office buildings, despite the fact that hey worked in stable as well as project teams. Four had chosen team/collaborative environments, two of them in a combination with activity settings. Two had chosen assigned workstations (permanent or periodic), two unassigned and two a combination of both. Home based telecommuting was used to a certain degree by almost all of the six organisations.

Becker and his co-workers at Cornell University also studied the organisational implications of different strategies when implementing new ways of working (Becker et al., 1994). Their research indicates that the most important aspect of implementing innovative workplace practices was the process behind the implementation. They used three pairs of dichotomies to categorise the implementation strategies:

- Business-driven vs. cost-driven strategies.
- Process-oriented vs. solution-oriented implementation.
- Strategic vs. independent initiative.

Business-driven projects are those whose starting point is an interest in exploring new of working. Cost reduction is typically not a major consideration here.

Cost-driven strategies are those whose primary motivation is the desire to reduce cost; that is that without significant pressures to reduce costs it is unlikely that the new workplace strategy will be implemented. The cost driven models at times seek business enhancement as well, but this is often a secondary and less important benefit for the project.

Process-orientated approaches develop a set of guiding principles and standardised methods for analysing work patterns and practices. They then use the information collected from these processes to develop custom-tailored solutions to each situation. Thus the workplace solution developed at sites across the organisation tend to be very different, even though the same principles and processes guide their development.

Solution-orientated projects identify a prototype workplace strategy and then work to implement the same strategy across many different sites, albeit with minor variations.

Becker concluded that business-driven and process-oriented approaches by far outperform their counterparts both concerning user satisfaction, work effectiveness, the life time of the project and acceptance throughout the organisation. Their implementation model has four phases:

(1) Reassess how and/or where work is done.
(2) Conduct fundamental changes in business practice.
(3) Devise alternative workplace strategies.
(4) Manage change process.

What the Cornell team found was that a cost-driven implementation process would typically exclude work reassessment and business practice change and focus on developing the alternative workplace strategy and associated space and technology configurations.

A business-driven strategy on the other hand would place much more emphasis on work reassessment and business practice changes. Business-driven implementation strategies often looked at the project as a means of re-engineering the organisation.

In business-driven strategies managing change which occurred as a result of the new way of working tended to be continuous, an actual education and training process after the implementation. Cost-driven strategies in comparison either excluded this change management phase, or used a lot of time after the implementation comforting and helping the employees to adapt to the new solution.

Table 8   Summary table of workplace case study characteristics.

Organisations	Workplace solution case	Additional data
ORG 1 Leading supplier of fertilisers, oil, gas, aluminium. About 40,000 employees, half of them abroad.	CASE 1-1: A pilot for the IT-division of ORG 1, comprising 50 employees from three different sections and the top management team of the division.	The organisation is planning to build a new HQ in the outskirts of Oslo in 3-5 years time. Altogether five pilot projects have been launched to gain experience with flexible work solutions and are being evaluated over a two year period.
ORG 2: Leading supplier of oil, gas and petrochemical products. About 17,000 employees, one third abroad.	Two pilot projects: CASE 2-1: The documentation centre of ORG2, (2-1.A: library and 2-1.B: IT based documentation), a total of 60 employees. CASE 2-2: The OD and change management group of ORG 2, a total of 15 employees.	The organisation started out in 1975 with open plan, landscape office solutions in their HQ. Since then, because of employee discontent, the organisation have rebuilt most of their premises into cellular offices. In 1997 the organisation started a process to introduce alternative workplace solutions. The first pilot project, for a project division comprising about 200 employees was turned down by the employees. The two pilot projects in our study are the second and third pilot projects in ORG 2.
ORG 3: Leading supplier of phones and services to private customers and small companies including mobile phoning and data communication. About 1.6 million mobile phones and 1.8 private phones subscribers.	Two divisions: CASE 3-1: The Private Phones Help Centre of ORG 3, a total of 75 employees. CASE 3-2: IT division of ORG 3, (3-2A: IT product & service development), a total of 400 employees.	In late 1997, two divisions of the largest telecommunication concern in Norway merged, and became ORG 3. In early 1998 the top management of ORG 3 created a co-location project, of which the cases both are part. In 2001, ORG 3 will move into a new HQ in the ouskirts of Oslo, being built by the telecommunication concern of which ORG 3 is a daughter company.
ORG 4: IT - company producing and selling electronic systems and services, and infrastructure services. Total of 1350 employees countrywide at the moment, but growing rapidly.	Two divisions: CASE 4-1: The electronic product & service development division of ORG 4, a total of 95 employees. CASE 4-2: The infrastructure division of ORG 4, (4-2A surveillance centre, 4-2B customer support) a total of 75 employees.	In early 1998 the new top manager of ORG 4 co-located about 500 employees from different divisions into a new HQ. The organisation is extending this HQ to include another 400 employees, ready for operation in 1999.
ORG 5: Norway's second largest financial concern, overall supplier of financial services. Total of 4300 employees countrywide.	CASE 5: The headquarter of ORG5, a total of 1440 employees.	In late 1998, ORG 5 co-located all its administrative functions and daughter companies into a new built HQ in central Oslo.
ORG 6: International OD firm. 3000 employees in 22 countries.	CASE 6: The headquarter of ORG 6, a total of 95 employees.	One of the first organisations in Norway to introduct alternative workplace solutions, implemented in 1996/97 by a new managing director.

The Organisations and the Workplace Solution Cases

Becker's model was used to describe the implementation model and to categorise the implementation strategies in the organisations studied. The participants in the study were leading Norwegian organisations with regard to size and position in their fields. Four of the six organisations in the study were either 100% owned by the state or by publicly authorities (ORG 2, 3 and 4) or had the state as a majority owner (ORG 1). Two were privately owned organisations (ORG 5 and 6). The characteristics of the organisations and the work place cases studied are summarised in Table 8.

Organisational Motives and Implementation Strategies

The perspective in this study was a top-management perspective and not a user perspective. The motives of the top management are summarised in Table 9. When facility or office managers have expressed additional goals, then these are also cited in the table, marked (FM).

Table 9   Summary table of management motives.

Organis-ation	Organisation objectives for introducing alternative workplace solutions	Initiators behind alternative workplace solutions Managment of implementation process
ORG 1	- Organisational development and change - Functional flexibility - Create a new image	Initiator: Human Resources manager on concern level. Implementation management: The pilot division.
ORG 2	- Added business value - Functional flexibility - Cost reductions (FM)	Initiator: The division managers. Implementation management: The pilot divisions.
ORG 3	- Organisational integration - Organisational development and change - Create a new image	Initiator: Human Resources manager of the organisation. Implementation management: HR department.
ORG 4	- Organisational integration - Organisational development and change - Functional flexibility - Create a new image	Initiator: The Human Resources and Administrative Functions managers of the organisation. Implementation management: The HR and AF.
ORG 5	- Financial flexibility - Functional flexibility - Cost reductions	Initiator: The Top management and the Board of the organisation. Implementation management (of alternative workplace solution): A "co-location" project group.
ORG 6	- Added business value - Functional flexibility - Cost reductions (FM)	Initiator: The top manager of the organisation. Implementation management: The top manager.

All of the organisations in the study were in the process of trying to develop more efficient and effective work-processes and at the same time changing rapidly. They were all dependent upon being able to recruit and keep the best young knowledge workers. All were in very competitive markets, with the objectives of almost all covering both dimensions which, according to Duffy, usually lie behind workplace innovations; adding value to business performance and driving down occupancy costs.

Table 10   Summary of implementation strategies.

Organisation	Implementation processes			
	PHASE 1 Reassessing how and/or where work is done	PHASE 2 Conduct fundamental changes in Business Practice	PHASE 3 Devise Alternative Workplace Strategies	PHASE 4 Manage change process
ORG 1	yes	no	yes	yes
ORG 2 - 1	yes	no	yes	no
ORG 2 - 2	yes	yes	yes	yes
ORG 3	yes	no	yes	yes
ORG 4	yes and no	no	yes	yes and no
ORG 5	yes and no	no	yes	no
ORG 6	yes	yes	yes	yes

In order to understand which of the two dimensions have weighed most, and thus be able to answer the questions asked at the outset of the study, the theories of Becker and his team were used (Becker et al., 1994). Their implementation model includes four phases, in an iterative process, as introduced earlier. When they analysed whether the implementation processes were business-driven or cost-driven they found that cost-driven processes would typically exclude phase 1 and 2 and focus on phase 3. Cost-driven processes also excluded phase 4 or used a lot of time after implementation helping employees adapt to the solution. Business-driven processes on the other hand placed much more attention on phase 1 and 2, and looked at the project as a means to re-engineer the organisation. Change management, phase 4, tended to be continuous. When this model was applied to the implementation processes in the Norwegian study, only ORG 6 and pilot 2-2 in ORG 2 qualify for being business-driven processes. The rest have conducted cost-driven processes in the terms of Becker and his team. The results are summarised in Table 10.

ORG 6 and pilot 2-2 in ORG 2 were the two organisations in the study where top managers reported a major improvement in their productivity and/or rate of return after they had introduced alternative workplace solutions, and that the improvement had lasted ever since (three and one year, respectively). They both reported that they used far less time on solving problems and implementing projects after they moved into in their new office surroundings and that this was the main reason for their improved productivity.

There may be other explanations than being business-driven in their approach for the achievement of the two organisations. Both had organisational development and change as their core business and hence knew more about business-driven change processes and team building than the other organisations in the study did. Being management consultants, they were also much more conscious about measuring the effects of change, and marketing their achievements than the other organisations in the study were.

Using Becker's terms 'process-oriented vs solution-oriented strategies' to characterise the implementation processes of the organisations in the Norwegian study, only ORG 6 and ORG 2 qualified for having had a process-oriented approach. The other organisations in the study used a solution-oriented approach in their implementation processes. Becker also found that business-oriented strategies as opposed to cost-driven strategies often resulted in significant changes in management philosophies and practices, work behaviours and attitudes, and culture. Based on interviews and observations, there is reason to believe that this is true for ORG 6 and pilot 2-2 in ORG 2 in the Norwegian study as well.

Conclusion

Do senior managers in Norwegian knowledge-based organisations have strategic objectives when introducing office change? Are alternative workplace solutions used as a strategic tool to support organisational change and development? Do top managers understand how alternative workplace solutions can be used as a strategic measure to add value to the organisation? Or is introducing alternative workplace solutions only a question about introducing more cost efficient workplace solutions, to build and to manage as a facility?

Yes, Norwegian senior managers undoubtedly do have strategic objectives with office change, and use alternative workplace solutions as a strategic tool to support organisational change and development.

It is more doubtful whether top managers really understand how alternative workplace solutions can be used as a strategic measure to add value to the organisation. If adding business value means having a business- and process-oriented approach to implementing alternative workplace solutions, only two organisations in this Norwegian study understood what it takes to achieve this.

The size of the projects may have played a role in this, however. In ORG 3, ORG 4 and ORG 5 between 500 and 1500 employees were being moved into new premises within a relatively short period of time. This reportedly called for simplification and universal solutions. ORG 6 and pilot 2-2 in ORG 2 in our study are relatively small and manageable in comparison (95 and 15 employees respectively).

Cost efficiency also may have been a decisive factor for choosing universal workplace settings. Process-based workplace solutions may result in higher productivity, but they may also hinder organisational flexibility, being tailor-made as they are to specific work-processes. They also demand a much higher degree of end-user involvement to succeed, which some top-managers find wasteful.

There may be a contradiction between the top management level and the operational level in organisations concerning workplace design, two forces pulling in opposite directions. Top management in large organisations may want universality in order to obtain maximum functional flexibility and cost efficiency. On the operational level however, department managers may need tailor-made solutions in order to obtain maximum productivity and competitiveness.

This is a dilemma, but also a challenge for work place designers and facility managers. As one of the interviewees said:

> "When industrial organisations design buildings for production, they always study their production processes, and how they can be improved, before they design the building they think is best suited to support the work processes. When organisations design, redesign, buy or rent an office building, they very seldom do that."
> (Project manager ORG 5.)

References

Becker F. D. Quinn K. L., Rappaport A. J. & Sims, W. R. (1994) *Implementing Innovative Workplaces: Organizational Implications of Different Strategies. Summary report*. Cornell University International Workplace Studies Program. Ithaca, NY.

18 Property Performance and Productivity

Nigel Oseland, Johnson Controls Ltd
Steve Willis, Sulzer Infra CBX Ltd

Property Performance Value

A property's performance can be assessed using many metrics, but ultimately its performance is a correlate of how well it supports the set of activities taking place within it. So it follows that office performance is related to how well it supports the occupants' normal work activities, and enhances their output and contribution to the business's overall performance.

Designing and managing workplaces in order to maximise staff performance has a cost associated with it and building performance must be assessed in this context of cost. 'Value for money', 'cost effectiveness', 'optimum efficiency' and 'maximising productivity' are the common language of businesses today, and reflects a shift from the image driven offices of the 1980s to the functional workplaces of the 1990s. These phrases all represent one common message: to maximise performance at minimum cost. For example, increasing productivity requires reducing costs with an increase, or no change, in performance, whereas reducing both costs and performance simply results in a low-cost low-quality solution.

Figure 26 The Performance Value Model.

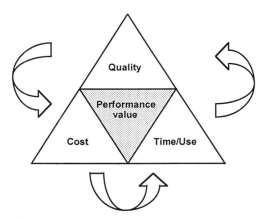

The key to success is to balance out costs with performance, but this requires evaluating both elements of the productivity equation - a task easier said than done! To assist in this a Performance Value Model has been developed, the three key elements of which are shown in Figure 26. The model is derived from the Total Quality Management (TQM) model, fused with the basic principles of project management, with the focus on: cost, time/use and quality.

The Performance Value Model is not only theoretical, it is a practical tool since all three elements are measurable and can be compared against industry benchmarks. Most organisations will be tracking at least one element of the model, usually cost, and some may have a handle on two of the elements.

Figure 27   Example of IPM Database Output.

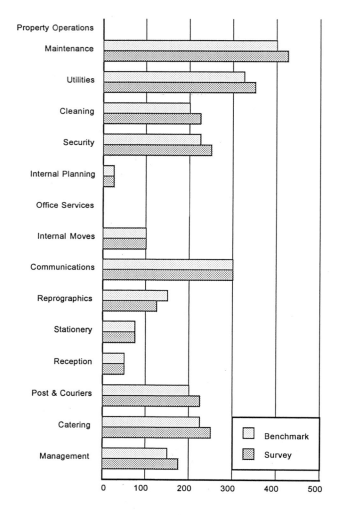

Johnson Controls regularly use all three elements to evaluate the holistic performance of workplaces and have created a database for each of the measures. The approach has been tried, tested and refined over several years. All databases of performance measures are dependent on the validity, reliability and accuracy of the data. The constraint over the accuracy of the data is the time and cost of collecting it. The Johnson Controls approach is to collect the data, for each element of the Performance Value Model, at a level of resolution which is practical and cost-effective. This level of resolution is lower than found in other databases dedicated to a single metric, but the interdependency between the cost, use and quality elements increases the overall accuracy and provides an internal self-test of reliability and validity.

Figure 28 Example of TUS Output.

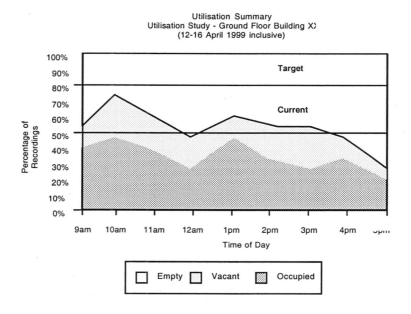

Figure 29 Satisfaction with Facilities.

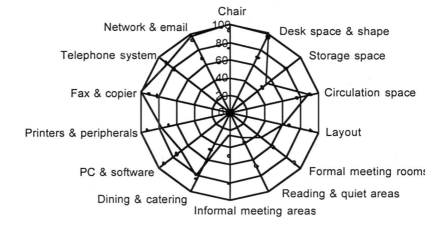

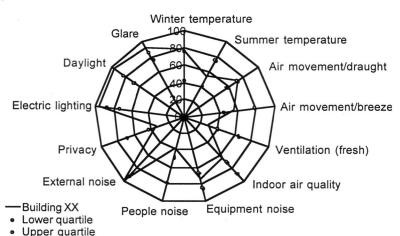

Figure 30 Satisfaction with Environment.

Three Metrics of Property Performance

'Cost' refers to the cost of the workplace facilities and building services. The International Performance Management (IPM) database is a knowledge-based system containing costs of hard and soft FM services for thousands of buildings located in 40 countries. A pro-forma check-list completed by the building FM provides the information required to benchmark the FM costs. The database allows FM costs to be compared with other similar buildings to ascertain whether the facilities offer relative value for money. Basic information of this kind, as shown in Figure 27, is required for developing a strategy to improve cost-effectiveness, but must also be considered in the context of use and quality.

'Time/use' refers to how efficiently the building space is being used over time. An under-utilised building, e.g. empty workstations and meeting rooms, is not cost-effective. Increasing occupancy rates, through desk-sharing schemes (e.g. team-share, hot–desking and hoteling) has been shown to reduce the amount of accommodation actually required to support the business. However, not all companies, or departments within a company, are suitable for desk-sharing. The first stage is to conduct a Time Utilisation Survey (TUS), as shown in Figure 28. This is an observation study in which the occupancy rates of workstations and rooms are tracked, along with an assessment of the activities carried out in the space. This survey and associated analysis predicts the level of mobility of the workers and the corresponding desk-sharing ratio which will maximise utilisation rates with minimum disruption to staff performance. The next stage is to use plans to ensure that the desk and room layout is space efficient and allows the flexibility required in today's team orientated organisations. Key performance measures derived from the TUS and plan analysis can be benchmarked against other buildings. Saved desk space can be used to produce function-driven work areas which support the range of activities taking place in the business e.g. workrooms, quiet areas, breakout spaces, and informal meeting areas.

Figure 31 Estimations of Downtime.

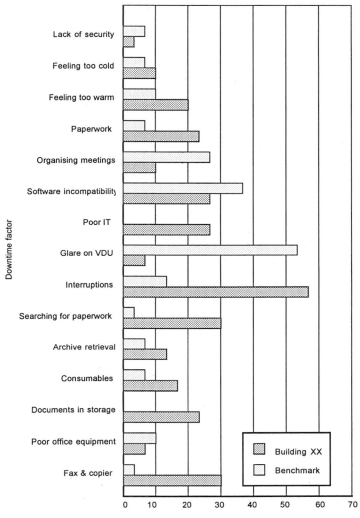

The third metric is 'quality', which refers to how well the office meets its intended purpose i.e. to support work activities and enhance worker performance. The first two metrics are 'hard' measures, being both objective and quantitative. Quality is more intangible and a 'soft' measure, based on subjective but nevertheless quantitative, data. Johnson Controls use a standard questionnaire survey, backed up by manager interviews and end-user focus groups, to establish the impact of the workplace on staff satisfaction and performance. Studies of workplace productivity (Oseland and Bartlett, 1999a, 1999b) indicate a strong correlation between performance and satisfaction. A workplace which maximises staff satisfaction is therefore one most likely to maximise performance. Sufficient surveys have now been carried out to produce a database of building quality. This allows a building to be benchmarked against the others so that its relative performance can be assessed.

Figures 29 and 30 show an example of how satisfactory a building's facilities and environmental services are in supporting the occupant's normal work activities. The building is compared with the best and worst performing buildings (upper and lower quartiles) in the database.

In addition to satisfaction questions concerning downtime were included. Here, building occupants found estimating the amount of work time that had been wasted, due to series of facilities issues, was an understandable and useful measure of performance for them. Small amounts of time wasted due to individual factors can amount to significant downtime, which is directly correlated with loss of earnings. Figure 31 indicates the type of results that can be obtained, but due to its subjectivity, caution is required in using and interpreting downtime data. However, when used in a comparative way, it can provide a key indicator of performance.

The questionnaire approach can provide a practical means of determining how well the property is performing overall in comparison to others, i.e. a 'health check'. This external benchmark is a means for visualising the shift to both the internal and industry standard of workplace quality, as shown in Figure 32. The survey approach also highlights the particular aspects of the facilities in most need of attention. This allows us, for example, to prioritise recommendations for workplace refurbishment in terms of quality and cost to develop a short-term 'quick hit' and long-term accommodation strategy. This will improve an organisation's internal quality standard.

Figure 32   Benchmarks Improve Quality.

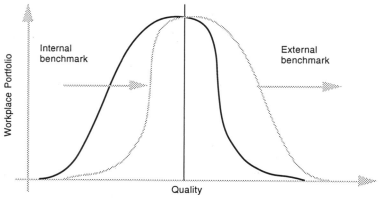

The Facilities Managers Role

According to BIFM (1999), "Facilities Management is the integration of multi-disciplinary activities within the built environment and the management of their impact upon people and the workplace". The normal focus for FMs is to aim to reduce office costs per square metre, but their role should be to provide productive workplaces, that is workplaces which maximise worker performance at minimal premises costs. The impact of 'facilities' on the whole organisation's performance should be considered, not simply the impact on the FM budget. The Performance Value Model offers a practical set of tools for assessing the value for money of facilities.

The first step to management is measurement, because if you can not measure performance then you can not be sure you are improving it over time. Undoubtedly, workplace performance is difficult to measure. However, staffing costs are around 75 to 85% of total business expenditure compared to 5 to 10% for premises (property and operating) costs. FMs can therefore not afford to ignore the impact of the office and its services on the output of their business, assume it is zero, or consider it negligible, when evaluating the value for money of their building. For example, small changes in employees' performance can have a significant effect on determining which refurbishment option generates the highest return for the organisation and in certain sectors, e.g. management and consultancy, just a 5% increase in staff performance can cover the cost of providing their accommodation (Oseland, 1999). The FM must be aware that what appears to be a low-cost building project can have a large hidden cost if it impacts on staff performance.

For FM to develop as a discipline, a paradigm shift is required with a move from focusing purely on premises costs to looking at holistic business performance. To increase productivity, a basic understanding is required of property and accommodation, building services, core business, motivation and management, organisational/cultural and personnel issues and of course finance. Using the tools that form the Performance Value Model can not only inform workplace design but also helps the change management required for successful implementation, by providing a highly visible approach which involves the staff at all stages.

The office-based organisation that looks at the contribution their office facilities can make to increasing staff performance will thus gain a competitive advantage over the business that only looks at providing poorer facilities at the lowest cost

References

British Institute of Facilities Management (BIFM) (1999) *What is facilities management?* Extract from BIFM web-site www.bifm.org/uk, May .

Oseland N. (1999) Don't be afraid of workplace productivity. *Workplace,* February.

Oseland N. & Bartlett P. (1999a) The bottom line benefits of workplace productivity evaluation. *Facilities Management Journal*, Vol. 7, No. 4, pp. 17-18.

Oseland N. & Bartlett P. (1999b) *Improving Office Productivity*. Addison, Wesley, Longman, Harlow, UK.

19 Corporate Real Estate Management

Ranko Bon, University of Reading

An annual survey of corporate real estate management practices has been conducted by the Corporate Real Estate Management Research Unit at the University of Reading since 1993. This Unit was established in 1991 with the objective of providing a firm and lasting foundation for the study of corporate real estate management (CREM) in both private and public organisations that are not primarily in the real estate business. Any organisation that occupies space is in the real estate business, and needs to manage it properly. So CREM covers the entire range of activities concerning buildings and land holdings held by an organisation, starting with investment and finance, through construction and facilities management, to reuse and disposition of property. This chapter summarises the results of six annual surveys from 1993-1998 and focuses on the incidence of corporate real estate management policies, functions, and activities, as well as the assessment of knowledge or skills relevant to the CREM function in the future. Both are of vital interest to educational institutions concerned with this field, as well as the personnel and training functions within organisations concerned with better management of their property. The surveys were conducted jointly with Johnson Controls Incorporated (JCI) since 1997, and with the International Development Research Council (IDRC) since 1999.

Survey Objectives and Scope

The main objective of the surveys were to facilitate inter-organisational comparisons, to help promote a more rapid diffusion of best CREM practices across Europe, North America and the globe [1]. Until 1997, the survey was aimed exclusively at chief real estate officers (CREOs), that is, top executives concerned with real estate in their organisations. From 1998 the survey covered CRE professionals in general.

Each year a number of CREOs from Europe and North America are asked to participate in the survey, which focuses on the incidence of CREM policies, functions, and activities across the Atlantic, as well as the assessment of the areas of knowledge and skills relevant to the CREM function in the future. The survey focuses on problems facing CREOs and other top-level executives concerned with the interaction between an organisation's property and its business activity. CREM employees (strategic concerns, financial orientation) do not normally include the personnel involved in facilities management (tactical concerns, organisational orientation) or building operation management (logistic or operational concerns, physical orientation). For further discussion of these three management profiles, (see Bon 1992, 1994b). The research that has been conducted to date suggests that top management is largely to blame for under-management of real estate portfolios, which form a significant proportion of most organisations' assets (Zeckhauser &Silverman, 1983 and Veale, 1989 and Avis et al., 1989).

Survey Design and Respondents

Both IDRC and the International Association of Corporate Real Estate Executives (NACORE International) have been instrumental in the distribution of the survey to their membership. The survey questionnaire includes some hundred questions divided into three parts:

- Background information on the corporate real estate organisation as a whole, and its CREM unit under the CREO's management.
- Corporate real estate management practices, including CREM policies, functions, and activities.
- Opinion survey regarding the role of corporate real estate in general and in the organisation in question, as well as the knowledge and skills crucial to CREM in the future.

This chapter focuses on only two sections of the survey: the CREM policies, functions, and activities in evidence over the study period, and the areas of CREM knowledge that are considered to be important for the future. Many of the organisations covered by the survey span both Europe and North America or are truly global in scope so that a clear distinction between the European and North American organisations is increasingly difficult to make and potentially misleading. Summaries of the relevant survey results are presented in Tables 11, 12 and 13. With the key findings highlighted in Figures 33-44.

Table 11   Respondents by region.

Number	1993	1994	1995	1996	1997	1998	Total (1)
Europe	15	24	11	8	18	17	93
North America	6	17	7	20	34	25	109
Total	21	41	18	28	52	42	202

Note 1: Total for all years include respondents who have participated in more than one annual survey

Table 11 shows the number of respondents over the six survey years. The participation rates from Europe and North America vary from year to year, but the total number of participants from the two regions is roughly the same over the period surveyed. The respondents are from major organisations, employing about 30,000 people housed in about 1,000 facilities on the average (Bon and Luck, 1999a). Again, most of them are international in scope, and many are global organisations.

Because of the variability of sample size, occasionally small sample sizes and considerable regional variability, the results of the survey must be treated with some caution. However, longitudinal comparisons of this kind often reveal credible and interesting trends. It is precisely the longitudinal aspect of this annual survey, which is of greatest value to CREM practitioners and to students.

Table 12   CREM policies, functions and activities.

Averages (percent) (1)	1993	1994	1995	1996	1997	1998
Property strategic plan(s)	90	88	78	82	83	89
Property MIS	86	95	89	82	88	85
Computer-based inventory	81	95	89	79	87	84
Separte evaluation of real estate	76	73	67	86	85	84
Performance benchmarking study(s)	52	78	72	68	67	82
In-house facilities management	52	76	67	75	64	76
Formal space standards	57	68	67	64	71	75
In-house construction management	29	61	44	71	67	69
Ongoing performance measurement	76	63	72	71	58	69
Disaster-recovery plan(s)	43	54	61	57	56	64
Policy on the use of consultants	67	85	67	64	65	62
In-house maintenance management	38	51	61	64	58	62
Property-by-property accounting	76	59	50	61	50	62
Internal rents	81	56	78	57	60	53
In-house design management	33	49	50	61	52	51
Desk-sharing policy	38	34	44	32	27	47
Teleworking policy	19	24	33	32	37	44
Clean-desk policy	14	22	22	25	17	22
Annual averages	56	63	62	63	61	66

Note 1: Ranked by annual ranking in the reverse order of years

CREM Policies, Functions and Activities

Table 12 shows the incidence of CREM policies, functions, and activities reported by the respondents. The results are ranked annually in the reverse order of years. If two items have the same ranking in one year, their overall ranking will depend on their relative ranking the previous year, and so on. The higher the ranking of a policy, function, or activity, the greater its incidence in the annual sample.

Figures 33 & 34.

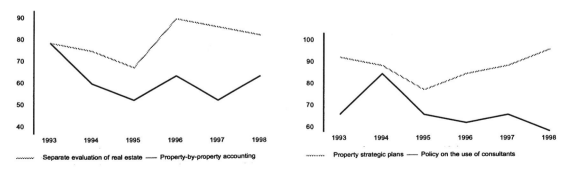

Zeckhauser and Silverman (1983) and Veale (1989) have identified several features that characterise organisations that successfully manage their real estate assets, with the separate evaluation of real estate and property-by-property accounting. Both are shown in Figure 33. Separate evaluation of real estate, which ensures that an organisation's main business and its real estate operations are not confounded in accounting terms, remains close to the top of the ranking. However, property-by-property accounting remains close to the bottom of the ranking throughout the period studied. This is surprising in view of the importance of real estate transactions, especially in the transition years.

Figure 34 highlights another aspect that is consistently high on the ranking list: property strategic plans, which are intended to bring into line the business or activity of an organisation and its property concerns. It is interesting to note the dip in the incidence of these plans in the mid-1990s. This was a 'transition' period during which many organisations restructured their property portfolios following the economic downturn. This finding 'correlates' with the policies on the use of consultants, which were most 'popular' during the transition years.

Figures 35 & 36

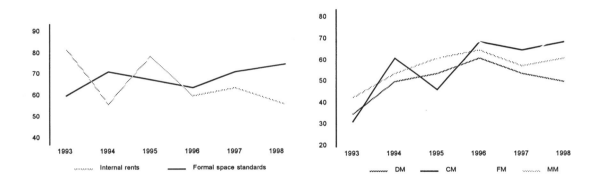

Figure 35 shows a gradual increase in the importance of formal space standards over the entire period of study, as well as a rapid decline of internal rents since 1995. The former trend is in line with the overall increase in the number of policies in this field, reflected in the annual averages in Table 2. In that light, the latter trend is somewhat surprising, however.

Figure 36 shows that many functions 'outsourced' in the early 1990s came back 'in-house' between 1993 and 1994. Bon and Luck (1998b) have analysed the behaviour of design management (DM), construction management (CM), facilities management (FM), and maintenance management (MM) functions and found that the shift away from outsourcing was relatively abrupt rather than gradual. There are also clear indications that the four functions are interrelated and that an organisation having one of them in-house is likely to have the others, also.

As can be seen in Figure 37, both the ongoing performance measurement and performance benchmarking have moved upward in the overall ranking between 1997 and 1998. Both activities had been declining between the mid-1990s and 1997, and the last finding is therefore quite encouraging. As argued elsewhere, these activities are crucial for the development of CREM into a bona fide management field (Bon, 1994b) and (Bon et al, 1994).

Viewed from the bottom up, the ranking of policies, functions, and activities is perhaps even more interesting. It is noteworthy in Figure 38 that the incidence of desk-sharing policies has jumped considerably between 1997 and 1998 following a decline since 1995. Also, it should be noted that more than 40 percent of organisations now have a teleworking policy. The incidence of the latter has gradually grown over the entire study period.

Figures 37 & 38

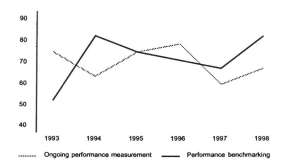

.... Ongoing performance measurement ——— Performance benchmarking

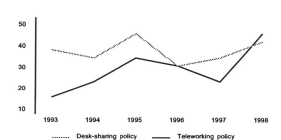

.... Desk-sharing policy ——— Teleworking policy

Figure 39 & 40

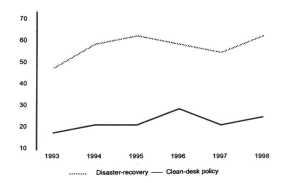

.... Disaster-recovery ——— Clean-desk policy

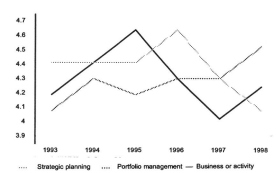

.... Strategic planning .... Portfolio management ——— Business or activity

The incidence of disaster-recovery plans and clean-desk policies is shown in Figure 39. Again, both show upward trends after some decline from the mid-1990s. The incidence of the former remains much greater than the incidence of the latter, however. In 1998 only about 20 percent of organisations have a clean-desk policy, which ensures that documents are easily retrieved after an explosion in the property or its vicinity.

Key Areas of Knowledge and Skills

Table 13 shows the opinions of the respondents about the relative importance of different areas of CREM knowledge and skills in the future. Perhaps this is the most important part of the CREMRU-JCI survey. Each year the respondents are asked to express their opinion on a 1 to 5 'least to most popular' scale. The results are ranked in the reverse order of years.

Table 13   Assessment of CREM knowledge & skills crucial in the future.

Averages (1)	1993	1994	1995	1996	1997	1998
Strategic planning	4.4	4.4	4.4	4.6	4.3	4.5
Business or activity	4.2	4.4	4.6	4.3	4.0	4.3
Portfolio management	4.1	4.3	4.2	4.3	4.3	4.1
Negotiation	4.2	4.1	3.9	4.5	4.1	4.1
Performance measurement	4.1	3.9	4.5	4.1	3.9	4.1
Total quality management	4.0	3.9	3.9	3.9	3.5	3.7
Environmental management	4.0	4.0	3.9	3.9	3.7	3.6
IT management	3.8	3.6	4.1	4.1	3.6	3.6
Workplace design	3.3	3.7	3.6	3.6	3.4	3.6
Facilities management	3.7	3.7	3.8	3.9	3.8	3.5
Personnel management	3.2	2.9	3.2	3.7	3.4	3.5
Real estate development	3.3	3.3	2.9	3.2	3.4	3.5
International finance	3.0	3.3	3.2	3.0	3.1	3.4
Contract management	3.2	3.3	3.3	3.6	3.4	3.3
Community relations	3.0	3.0	3.2	3.4	3.0	3.3
Government regulation	3.3	3.3	3.6	3.9	3.2	3.2
Security and safety management	3.1	3.4	3.5	3.1	3.2	3.0
Tax management	3.1	3.0	3.2	3.0	2.6	2.7
Annual averages	3.6	3.6	3.7	3.8	3.6	3.6

Note 1: Ranked by annual ranking in the reverse order of years
(1 = least important; 5 = most important)

As can be seen from Figure 40, strategic planning has not only emerged on top, but also has received a strong boost between 1997 and 1998. In this last period, portfolio management is eclipsed by the organisation's business or activity, the importance of which had declined between 1995 and 1997.

Figures 41 & 42

Performance measurement and total quality management (TQM), which were declining in importance between 1995 and 1997, have also shifted upwards between 1997 and 1998. This is shown in Figure 41. Again, this change is encouraging, because both activities are central do the development of CREM as a field.

As Figure 42 testifies, facilities management and workplace design are consistently ranked relatively low. However, the importance of workplace design has increased between 1997 and 1998, whereas the decline in importance of facilities management has continued since 1996.

Trends in environmental management and IT management are of particular interest, as shown in Figure 43. The importance of environmental management has steadily declined throughout the study period. The importance of IT management rose in 1995 and 1996 but it has declined since. The business systems required for effective CREM seem to be already in place.

Figures 43 & 44

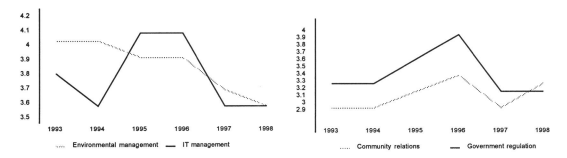

Figure 44 shows the perceived importance of community relations and government regulation. Both are considered relatively unimportant, in spite of the fact that they are rather likely to increase in importance in the future.

This overview suggests a few generalisations. Most of the areas of knowledge and skills that emerge close to the top of the ranking are a part of business administration in general. However, some of these areas are not currently emphasised or even covered by programmes in business administration. This suggests a need for programmes with an emphasis on CREM as such, in which the general principles of business administration can be applied to the specifics of this field. In this context, it is perhaps not surprising that several educational programmes in CREM have emerged in recent years [2].

It can be expected that an important part of education in CREM will remain in the form of short or modular programmes for those with specific needs and interests, but it is conceivable that some of the programmes on offer will in time attract students without prior experience in the field. For the time being, education in CREM is still largely in the domain of relevant professional organisations. Over time, stronger interaction between these organisations and universities can be expected, as well.

Conclusions

This chapter has only focused on some of the most interesting findings with a two-fold emphasis on the incidence of CREM policies, functions and activities, on the one hand, and on knowledge and skills relevant to the CREM function in the future, on the other. This information is of central importance first to educational institutions on both sides of the Atlantic, second to personnel departments in organisations concerned with better management of their real estate. This includes the educational concerns of international organisations, such as IDRC and NACORE International, which play increasingly important roles in the education of their members. The educational effort in this field has intensified throughout the 1990s, indicating its growing maturity. The CREMRU-JCI survey hopefully offers a useful tool in educational efforts in this area and the continuation of the survey and its internationalisation will lead to a better understanding of the future as it unfolds.

Notes

(1) The results of the survey since 1993 were presented by Bon (1994a, 1995, 1997) and Bon and Luck (1998, 1999a,b). This paper is based on an earlier one by Bon and Luck (1999a). For more information about CREM see, for example, Bon (1992, 1994b), which are based on Bon (1989: Chap. 5).

(2) There are Masters programs at the University of Reading, South Bank University (London), City University (London), and the European Business School (Frankfurt). In addition, the University of Delft, IDRC, and NACORE International offer professional development programmes.

(3) Barry Varcoe, Director of Strategic Development at Johnson Controls Inc., must be thanked for his generous support of the survey and for his contribution to the joint venture between the academia and industry.

References

Avis M. V. Gibson & Watts J. (1989), *Managing Operational Property Assets*, Department of Land Management and Development, University of Reading, Reading.

Bon R. (1989), *Building as an Economic Process: An Introduction to Building Economics*, Englewood Cliffs, New Jersey: Prentice-Hall.

Bon R. (1992) Corporate Real Estate Management, *Facilities*, Vol. 10, No. 12, December, pp. 13-17.

Bon R. (1994a), Corporate Real Estate Management in Europe and the US *Facilities*, Vol. 12, No. 3, March, pp. 17-20.

Bon R. (1994b) Ten Principles of Corporate Real Estate Management, *Facilities*, Vol. 12, No. 5, May, pp. 9-10.

Bon R. (1995) Corporate Real Estate Management Practices in Europe and the US: 1993 and 1994 Surveys, *Facilities*, Vol. 13, No. 7, June, pp. 10-16.

Bon R. (1997) Corporate Real Estate Management Practices in Europe and North America: 1993-1996, *Facilities*, Vol. 15, Nos. 5-6, May-June, pp. 118-124.

Bon R. & Luck R. (1998) Annual CREMRU-JCI Survey of Corporate Real Estate Management Practices in Europe and North America: 1993-1997, *Facilities*, Vol. 16, Nos. March-April, pp. 82-92.

Bon R. & Luck R. (1999a) Annual CREMRU-JCI Survey of Corporate Real Estate Management Practices in Europe and North America: 1993-1998, *Facilities*, Vol. 17, Nos. 5-6, March-April, pp. 167-176.

Bon R. & R. Luck (1999b) Outsourcing of Property-Related Management Functions in Europe and North America, 1993-1998: A Note, *Construction Management and Economics*, forthcoming.

Bon R., McMahan J., & Carder P. (1994) Property Performance Measurement: From Theory to Management Practice, *Facilities*, Vol. 12, No. 12, November, pp. 18-24.

Veale P. R. (1989) Managing Corporate Real Estate Assets: Current Executive Attitudes and Prospects for an Emergent Management Discipline, *The Journal of Real Estate Research*, Vol. 4, No. 3, Fall, pp. 1-22.

Zeckhauser S. & R. Silverman (1983) Rediscover Your Company's Real Estate, *Harvard Business Review*, January-February, pp. 111-117.

20 Adaptation and Sustainability

David Kincaid, University College London

The final ten to fifteen years of the last century has seen the beginnings of what, this chapter will argue, may be an important change in the use of buildings and infrastructure. This arises because of developments in information technology (IT) and the effect this has had on how work is done. This effect is heightened by the tightening of the environmental agenda and the effect of this on energy use and the methods of production of the outputs of economic activity. Early evidence of this change appeared as organisations in both public and private sectors found that staff were able to carry out office type work in a number of different settings through linked computers, facsimile machines and other IT communications mediums (Becker et al., 1991) These capabilities led to experiments with 'home-working' (APR et al., 1992) and other flexible time/space ideas. These, in turn, influenced thinking aimed at lessening environmental impacts by reducing transportation needs. From these developments, at least in part, new life style preferences, characterised by looser ties between individuals and organisations, emerged as recounted by authors such as Charles Handy (1991) and his subsequent works. When brought together all of these changes have had the effect, for many organisations, of both reducing their demand for office space and altering the nature of that demand physically and tenurally.

In manufacturing similar IT systems control the robots of the new flexible manufacturing processes in their smaller and often 'dark' people free spaces. An example of this was IBM's Personal Computer line in Greenock Scotland which by the early 90's assembled all of IBM's PC's for Europe, the Middle East and Africa in a space of about 2500 square metres. In this space less than 10 highly skilled engineers monitored the control computers and robots on the floor. Similar circumstances apply in Vauxhall's engine manufacturing facility in the English Midlands where robots produce one complete range of engines in space of a comparable size. Similar developments can be found in healthcare as the hospital sector finds the burden of post-operative care is reduced by the increasing use of micro surgery. Again, the effect is to both reduce and alter the characteristics of the requirements placed on the built environment.

The chapter will explore the present and possible future extent of these changes in demand and will seek to place these in the context of the sustainability agenda. This will be done through examining how existing buildings may be adapted to meet these new needs and how new buildings may be designed to allow sustainable adaptability to meet future needs. It seems evident from an examination of publications in this area that there is very little hard evidence available on the extent and consequences of many of the technological developments discussed. Accordingly the chapter will also consider the direction policy might take in order to develop more sustainable cities, both in relation to regulation and the research needed to support the continuing development of policy.

Physical Aspects of Sustainability

Before exploring further how to weave together aspects of the changing demand for buildings and how this interacts with the regulation and production of the built environment, a brief consideration of the physical sustainability objectives may help to clarify how future potentials could be achieved. From a physical standpoint sustainablility for the built environment is concerned with the level at which energy transformation, material extraction and eco-system impact can be allowed to occur in perpetuity in the creation and use of buildings and infrastructure.

Questions relating to the sustainability of the activities arising from the use of this built environment and the natural environment are of course at least equally important but are not addressed here. Even for this subset of the larger sustainability question the issues are complex and difficult. In this environment most policy makers, engineers and scientists take the view that the least we can do is to limit the damage while achieving worthwhile objectives to support the well being and prosperity of our communities. This has led to a number of regulatory, often formulaic, prescriptions for energy use and materials application emerging within the UK. These are concerned with items such as wall thermal conductance or environmental scoring systems for buildings (BRE, 1993). Other work has been done in the UK to design and build large scale buildings which are naturally ventilated or use recycled materials (Happy, 1997). Those, particularly within academia, concerned with finding a direct way through the complexity of these issues have sought to develop or apply theoretical models to issues such as transport and patterns of growth and development (Steadman et al., 1998).

All of these activities are worthwhile but merely scratch the surface of the sustainability issue primarily because they affect so little of the built environment we use. This is because only 1 to 2% of the total building stock of the UK is made up of new works in a typical year (DOE, 1987). The percentage of new works is even lower for roads and streets and lower still for railways. This inevitably means that we have to look to what can be done with what we already have if we are to significantly benefit the sustainability agenda over the next 20 years. But what can be done with existing buildings and infrastructure must, at least in part, depend on the nature and extent of the activities which the built environment serves. As was pointed out earlier, there is clear evidence that the demand side of the built environment equation may be significantly changing in both quantitative and qualitative ways. In any event by not examining the demand side of the equation we may be putting ourselves in the position of mere spectators who see the output of economic activity while having no understanding of what is necessary to the input. This a poor position from which to make policy and a near impossible position from which to choose the particulars of the physical makeup and configuration of built environment.

Before exploring the demand side further it is worth noting that there are at least certain aspects of the built environment which provide reason to regard sustainability as a realistic prospect. Many structures have survived over hundreds and in some cases thousands of years; the medieval Cathedrals of Europe and the aqueducts of Rome are obvious examples. As my colleague, Professor Bev Nutt has commented, sustainability is often merely a matter of planning and designing for the long term, rather than the short and medium. If the materials used in the first place are robust and the value placed on the maintenance of the material and the spaces created or the functions provided by the entity are worthwhile then there is good expectation of at least very extended life capability. This may approximate sustainability, particularly if the building or structure can be adapted to a whole range of human uses as discussed later. Can this longevity be extended beyond assemblies of stone, concrete and glass to the pipes, wires and fixings of contemporary built environment? Perhaps it can at least in the form of reusable materials, though with some expenditure of energy. Is this really back to saving old string? These questions can only be looked at realistically in the context of the uses to which the built environment is put.

Changing Use Demands

Economists identify over 500 generic economic activities in the Standard Industrial Classification system used world-wide. This forms the framework for all measurements used by government to inform economic policy and practice. Most of these activities make at least some use of buildings and infrastructure but despite this no attempt is made to measure any aspect of these uses insofar as they impact the built environment except for the purposes of statutory planning. Planning activities do provide data on use but only within 17 much broader categories of use activity. Accordingly we are not well informed on the detail of the specific type of work that is done within buildings located variously and having many different physical characteristics. It is then perhaps not surprising that when there is a step change in how we carry out our activities within an identified field of economic activity we cannot measure the impact that this might have on the demand for space or function. The sustainability agenda may however oblige us to improve our information on and understanding of the fine grain detail of the EIC activities if we are to understand how best to use what we have and the extent to which we need to adapt, extend or demolish our buildings and infrastructure. But would the undoubtedly costly assembly of so much detail make a worthwhile contribution to policy or design? A partial answer to this question may lie in looking more closely at some of the more evident examples of demand side changes. If whole sectors of economic activity are shedding buildings in substantial numbers due to demand changes then perhaps a closer examination of demand detail is warranted.

Examples can be found by looking at the recent developments in organisations such as British Telecom (BT), International Business Machines (IBM) and Surrey County Council in the UK all of which organisations had specific programmes for reducing space demand and encouraging more flexibility in work setting choices for staff. The effects within BT emerged during the course of Research into Refurbishment for Change of Use (Nutt et al., 1996). BT were Industrial Partners in the research and provided data on developments in their own portfolio which was very instructive. They were at that time looking to dispose of ninety million square feet (836,000 sq.m.) of building floor space (more than 10 Canary Wharf's) while at the same time expanding their telecommunications business. They wished to retain only about ten million square feet in 2010. Much of the disposal (40%) was related to the obsolescence of telephone exchange buildings no longer needing to house switching equipment as very much smaller computers do the job on microchips. That part of the disposal was no different from closing a factory except that the equivalent of the factory remains functioning. Perhaps even more interestingly, most of the remaining disposal related to the closure of offices, as the numbers of people needed to run the business reduced dramatically.

IBM in the UK found that when in the mid 1980s they introduced a network system linking computers on every employees desk in all of their 70+ buildings spread throughout the UK, the numbers of employees simultaneously using their assigned office space declined. Surveys in 1989 and 1990 showed that this decline represented, for some groups, usages as low as 35% and often in the 50% to 60% range (DEGW, 1998-90). All this was happening as business was growing by nearly 20% each year and staff numbers were constant. A programme to reduce desk provision and save space was taken in the early 1990s under the banner of the SMART programme (Space, Morale and Remote Technology) and led to some of the first installations of Group Address Facilities in the UK. Professor Frank Becker at Cornell reported on this programme and several similar programmes around the world in his report Managing Space Efficiently (Becker et al., 1991).

Sceptics will at this point be tempted to dismiss these cases as relating exclusively to the 'high tech' private sector. However, this would ignore the widespread initiatives found in Local Government and reported on by Edwin O'Donnell in Report for his MSc in Facility Management at UCL (O'Donnell, 1996). Working in Surrey County Council himself and leading teams involved in New Ways of Working O'Donnell was able to access considerable data on the extent to which other Local Authorities were responding to the capabilities of new Information Technologies and new attitudes to work to refashion their use of the buildings and reduce significantly the amount of space needed. It would appear from this evidence that Local Authorities provide a considerable test bed for experiments in new ways of working. The full extent of this activity has never been measured so it is impossible at this stage to size the effect of this beyond saying that it can be found in many UK Local Authorities.

From these illustration it seems apparent that significant changes have already occurred in many organisations. By extension there may be considerable scope for similar effects in both the public and private sector as IT impacts become more ubiquitous and IT itself develops further. It is already evident that there is a growing belief that the impact on retailing activities will be profound. The current behaviour of the equity markets in relation to Internet book retailers such as Amazon the book retailers is an indication of this view. If the speculators are right perhaps existing city and town centres will serve as places for a type of 'recreational shopping' while the edge of town supermarkets become warehouses for the bulk of routine domestic purchasing. Such a change may be speculative but it is now feasible. If it occurs then again demand will have a profound effect on achieve a sustainable future.

Before leaving the subject of demand it is worth noting that all the above cases have the potential to affect demand factors for housing. Thus, as was found by BT in their studies of 'home working', there is a gradual growth in the extent to which people choose to work at home, at least some of their working week. The pattern of this kind of working is variable and somewhat unpredictable, but when it does occur it naturally has an effect on the built environment. At its simplest working at home creates a need for a separate room in a dwelling to cater for the work activity, though it may simply create a second purpose for an existing room. It can increase energy demands in the domestic sector and it has led already to an increase in the number of second telephone lines installed in housing (Financial Times, 1999). In some ways this is a reversion to pre-industrial cottage industry pattern and as such could have a profound effect on what is regarded as normal use acceptable within residential areas. Many have found this way of working disagreeable for social reasons or impractical for their type of work so it is unlikely to be anything but a part time feature of life for many and a full time feature for a few. It is however a real change in demand affecting housing and transport at present to an unmeasured degree.

The issues discussed above suggest that there may be much to be learned from a close look at changing use demands. This leaves the question of what can feasibly be done with the existing building stock to adapt to these demands.

Aspects of Adaptability

The research into Refurbishment for Change of Use (RCU) referred to earlier, provided a useful insight into the extent and nature of adaptive reuse activities occurring in London in the mid nineties. Examination of origin and destination figures at that time showed clearly that the dominant trend was to convert older or badly located office space into housing, as illustrated in Figure 45. This trend could also be found in a number of other cities particularly in North America. Because the cost of office space was often closer to the cost of housing the pace and extent of activity was greater affecting even high quality office. At that time it was also evident that there was a strong trend towards change of use from industrial to housing. Resistance to change from industrial use by planners and planning committees was however seen to be very much stronger. Thus derelict industrial space in the UK has been left for many years as determinist authorities attempted to insist on achieving employment targets for their areas (Marsh, 1994). Notwithstanding this resistance some of the most significant redevelopments in London, as elsewhere, have been associated with the adaptive reuse of manufacturing, warehouse and market trading building complexes. Examples of this are seen in the conversion of the Bryant and May match factory in Bow London (once the world's largest factory building) into housing, the conversion of the Butler's Wharf warehouses on the South Bank of the Thames into housing (Nutt et al., 1996) and retail and the adaptation of Spitalfields flower and vegetable market into restaurants, shops, recreation, music venues and other uses.

Figure 45 Origin and Destination Uses of Planning Applications involving Change of Use.

		Destination Uses					
		Residential	Retail	Industrial	Office	Other	
Original Uses	Residential	3.5%	1.2%	0.0%	3.2%	0.6%	8.5%
	Retail	1.1%	1.2%	0.3%	2.6%	0.6%	5.8%
	Industrial	7.6%	0.5%	1.5%	3.5%	2.3%	15.4%
	Office	33.7%	4.7%	0.2%	1.2%	9.6%	49.4%
	Other	10.8%	0.8%	0.3%	5.5%	3.5%	20.9%
		56.7%	8.4%	2.3%	16.0%	16.6%	100%

The RCU research considered how to deal with the question of what use a particular building could be best adapted to when found obsolete for its previous use (Kincaid, 1997). Leaving the financial analysis to be considered as a second stage decision, this led to the development of a computer based 'Use Comparator' which takes account of the physical and locational characteristics of a building (the supply side) and compares these with a set of such characteristics which describe the needs of a particular usage (the demand). In order to create the Comparator it was necessary to identify from the Standard Industrial Classification (SIC) list of over 500 economic activities those activities that could be reliably given a set of characteristics. From this exercise 76 generic uses were characterised. The computer then indicates the likely best uses for a given building once it has been characterised and can readily provide more alternatives relating to characteristics changes that selective demolition can introduce.

The identification of the characteristics most appropriate to 76 different uses of course suggests that designers could, by this means, take account in both new work and refurbishments, of those characteristics that give the building the most robust set of possible use options for the future. This could in itself make a significant contribution to the sustainability of cities as they are developed and changed. Discovering this connection between design and use characteristics may be the first step in developing a more informed approach to how we should regulate in this area as well though it will be essential to allow redundancy and flexibility to incorporate the uncertain technologies of the future.

Adapting to Changing Demands to Achieve Sustainability

How then can use be made of what is known about changing demands and the adaptive capability of buildings and infrastructure to inform the developing understanding of sustainability? Quantitative changes in demand are clearly suggested by the examples given above which mainly indicated reductions in non-residential and increases in residential demand could occur. Qualitative changes are evident as well but the adaptation of buildings is a mature activity and decisions can be well informed by using the techniques tested in the 'Use Comparator'. Dramatic changes such as might occur in retail could provide a lifeline to town centres and transport without heavy regulatory intervention. Such an optimistic outlook however ignores the history of seemingly inexorable growth of urban areas and private transport which challenge any notion of true sustainability.

As with so many other sustainability initiatives then perhaps all that can be done is to incrementally improve environmental performance by avoiding those actions that lead to obviously damaging consequences and by encouraging those actions that at least minimise the damage. In this vein a number of commentaries may be helpful.

- **Redundancy**: In the first place we should respond to the uncertainty of future qualitative demands by providing some redundancy in what we build new or change. Too much floor to floor clearance is wasteful in both the long and short term; too little is always wasteful in the long term as use changes. A generous clearance for a range of possible uses serves both the long and short term. Similar commentary can be made about strength, depth, height and size. Materials too should be selected to allow for these balances in benefit over the short and long term.
- **Ambiguity**: A variety of possible uses, not bound by the contrived categories of regulation, should be assumed likely for a building. Nothing should be done to unduly constrain the adaptation of a building to a range of future uses. Equally a single easily defined use for a building should be avoided. Policy makers in the UK are now beginning to recognise the importance of this factor.
- **Flexibility**: The most sustainable way to provide flexibility may well be to create spaces within a building which have such a quality that people adapt their activities to suit the building and not the building to match their activities. When this possibility is exhausted a building should be adapt-

able through its geometry, fabric and structure (in most cases) without the need to reinvent its essential morphology nor should component cellularity constrain it's ability to adapt to new technologies.

- **Constraint**: There are few physical constraints to building anything anywhere for any purpose provided the resources can be found. Thus most constraints in the built environment relate to financial issues and regulation. In both cases excess can damage sustainability . Too much finance leads to physical excess in size, materials and energy and must be constrained by regulation. Too much regulation leads to 'heritage blight' as adaptation is arrested and cannot be constrained except through political intervention.

- **Design:** It is most unlikely that sustainability can be achieved merely be the constraints of regulation. Regulation is essential in public affairs but is not the basis for producing food or manufacturing products. These require management, design and operations to meet the needs of modern societies. Why then should we expect sustainable built environments to emerge simply by creating regulatory frameworks? To achieve this we need to decide what to keep, what to build, what material to use for which building and how to meet the changing demand requirements of the users. That complex activity is largely what may be called design. Sustainability of the built environment can only be approached through informed design to meet the ever changing demands of the users of that environment. This will have to done for the individual buildings and infrastructure as well as for the aggregation of these into the totality of towns and cities. Anything less is bound to fail the stringencies of sustainability.

Policy Options for Physical Sustainability

To achieve even these first steps towards sustainability will require that Government play an active role in both regulation of the behaviour of companies and individuals but also in expanding our knowledge and understanding of the demand side of our use of the built environment. This will require policy initiatives in two spheres in particular.

Research

(1) A much better understanding of what activities are carried out by organisations and individuals in buildings is required. New technologies have transformed this use both qualitatively and quantitatively and yet very little funding is available from Government to investigate the nature and extent of this change and to extrapolate from this to better understand future possibilities. New ways of working are becoming well established but no information on how this affects use is available.

(2) The further development of our understanding of the characteristics of buildings that are best suited to different generic types of uses is essential if we are to understand how to design and change our buildings to achieve sustainability. Despite years of work in looking at the material fabric and structure of buildings, hardly anything has been done to look at how this relates to use.

Regulation

(1) Considerable long term benefit would flow from a planning requirement that all permissions should include a demonstration that proposals allow for a range of uses in future, in most cases, or at least a demonstration that an appropriate range of future uses is not precluded by a particular design solution. This should apply to both any new building or an adaptation.

(2) A set of clear and unambiguous criteria should be established so that a building which is deemed to be a 'heritage building' can be demonstrated to be such. One part of that criteria, for most buildings in this category, should be that the building is adaptable to a recognised set of uses without loss of its essential character.

(3) Building regulations must move towards a set of re-usability or recycle-ability criteria for construction works whether related to new or refurbishment activities.

It is unlikely that these kinds of proposals which seek to emphasise the importance of the demand side of the equation and the capability buildings and infrastructure have for adaptation will on their own move us into a world which achieves sustainability. However they are an essential component of the sustainability agenda, in the absence of which we will be unable to move successfully towards this elusive goal.

References

APR et al. (1992) *The Home Office Report*, DETR, London.

Becker F. et al. (1991) *Managing Space Efficiently*, Cornell University NY.

Building Research Establishment (1993) *BREEAM Guidelines*, BRE Watford.

DEGW (1989-1990) Internal Consultancy Reports to IBM, *unpublished,* London.

Department of the Environment (1987) *Re-using Redundant Buildings*, London.

Financial Times Report, 16 February 1999.

Handy C. (1991) *The Age of Unreason*, Random Century, London.

Happy J. (1997) *Green Buildings*, MSc Report, UCL, London.

Kincaid D. (1997) *Managing Adaptive Reuse of Buildings*, BIFM Conference Proceedings, Cambridge.

Marsh G. (1994) *The Home Office Report*, DOE, London.

Nutt B., Kincaid D. & McLennan P. (1996) *Refurbishment for Change of Use*, Final Report to the EPSRC, UCL, London.

O'Donnell E. (1996) *Flexible Working Practices in Local Government Offices*, MSc Report, UCL, London.

Steadman P. (1998) *An Integrated Buildings, Transport and Energy Model of Swindon*, EPSRC, Swindon.

Part Four

The Knowledge Trail

21 Introduction to the Knowledge Trail

Information and Communication Technology (ICT) is altering the ways in which business is conducted, fundamentally. The functions and operations of facility management are also bound up inevitably in this ICT revolution. This technology can make an instant impact on some areas of potential application, particularly those with well recognised product lines, those providing simple routine services, with stable data fields, and those areas that are supported by well tested methods, techniques and procedures. So in FM, e-commerce systems for the procurement of standard items and basic services are already becoming available. Applications in areas with secure information are following fast, for example, with systems for managing legislative compliance and life-cycle maintenance. But in regard to many of the functions, operations and services of FM, ICT technology may have to be put 'on hold' until FM develops specific expertise, methods and techniques, unique to its own field.

The FM Knowledge Trail in the 'year 2000', starts from a position that relies largely on borrowed management concepts on the one hand, and on imported technical expertise from other professional fields of activity, on the other. While the relevancy and potential value of available technical and management expertise is recognised, its application to the specifics of facilities operations and management is poorly developed. So the FM Knowledge Trail is at an early stage of development in which:

- It sets out from an ever widening and ill-defined sphere of activity.
- It still needs greater internal coherence for many working in the field.
- It lacks external coherence to many corporate and business organisations, and to the educated public at large.
- It has too few secure methods of its own to underpin good practice.
- It has hardly begun to make its own distinctive contribution within the management field.
- It is insufficiently supported by an adequate knowledge base.

If FM it is to benefit from the ICT revolution over the next few years, then it must innovate and act to build its own distinctive knowledge base with supporting application protocols. Two knowledge perspectives will have to be considered, those of the corporate organisation and those of the individual employee. Knowledge within the corporate context, can be more valuable than the goods or services that an organisation provides. The knowledge content of many goods is now more valuable than the product itself (Stewart, 1997). But what of the value of FM knowledge? Could it also become an important component of business value? For example, the development of company intranet knowledge systems for business support such as those in Skandia, Ericsson and Andersen Consulting are well known (Edvinsson et al., 1997) and (Baladi, 1999). The degree to which these knowledge systems have, in the past, been developed to incorporate facilities management functions is less clear. New strategic knowledge for FM could become a principal component of corporate knowledge value, as information begins to define market share for products and services around the world, with ICT providing the means for 'real' time management, globally.

Similarly initiatives for innovations in the individual's FM knowledge systems are hard to find. If the changes to work and employment practices, as detailed in Part Two, continue with more flexible employment contracts, project based working and smaller companies generally, then the need for self contained, personal, versatile and portable FM knowledge systems will intensify. Personal knowledge systems will need to be compatible, on the one hand. with corporate systems under network or partnership arrangements, while on the other hand, they must continue to augment an individual's personal knowledge system over time, shedding redundant knowledge as appropriate. They must also be capable of flexibility adjustment to meet the specific demands of different projects. Understanding the types of knowledge facility managers may need, use and create in the future is an important area for investigation. But where will the new corporate and personal FM knowledge that is needed to run these systems come from? Typically innovations originate from any of four sources:

- **Operational Origins**: Practice-led innovations where those with direct practical experience of current FM tasks and operations find ways in which existing best practice might be modified or superseded by better working processes, methods and procedures.
- **Problem Origins**: Theory-led innovations where improved understanding of the fundamental nature of recurrent FM problems and phenomena could give rise to novel developments and new techniques.
- **Personal Origins**: Innovations based on new FM ideas and concepts, where personal insights, imagination and creativity could lead to unique FM developments, services, products and processes.
- **Contextual Origins**: FM Innovations triggered by external rather than internal factors, particularly by global developments, political and business policies, technological advance, market competition and fashion.

These four sources of innovation occur in combination, but varying with the field of application. For example, throughout the last century progress in science and engineering, and the inventions and patents that resulted, tended to arise out of 'operational' and 'problem' origins. In contrast innovations in architectural design have, in the main, relied on 'personal' and 'contextual' origins for new ideas. Innovations in the field of management are more problematic, with some friction between general business management's reliance on 'personal and contextual' origins for ideas, and management science's emphasis on 'problem and operational' priorities. In the case of facilities management a balanced approach seems to be required due to the broad nature of the FM field, straddling as it does, the concerns of business and property, addressing both management and technical issues.

Most knowledge management systems face a common problem in reconciling the 'demand' for information with its 'supply'. On the demand side, knowledge requirements arise under a particular set of circumstances and tend to be for 'situation specific' information within an 'organisationally specific' context. The requirements of each knowledge customer are likely to be somewhat 'unique'. On the supply side, particularly in the case of electronic information systems, the available knowledge will consist of 'generic information' and 'generic experience' only. Knowledge will be 'general' and 'non-specific' to the needs of any individual client.

Situation specific information will relate, therefore to the particular requirements of an individual organisation under its current circumstances. Typically in FM, these requirements are 'case specific' and 'practice-led'. Here, the assessment of demand, the search for and access to appropriate information, will be driven by the expectation of subsequent improvement for the short term benefit of the specific organisation and its staff.

In contrast, generic knowledge will relate to recurrent problem issues that are seen to be fundamental to and shared by all in the FM field. The development of generic knowledge bases needs to be structured around a framework of 'practical -theory' that defines the scope, nature and characteristics of the recurrent problems that are faced. The acquisition of generic knowledge will normally involve systematic investigation, supported by reliable data, with results validated prior to implementation. Generic knowledge, by definition, should be capable of general and wide implementation. It must not be expected to be able to resolve the specific problems of any particular organisation and its circumstances directly.

In the field of facility management this fundamental difference between generic knowledge and case specific information is rarely appreciated. On one hand, many FM practitioners hold unrealistic expectations of the benefits that research results and the knowledge gained, might hold for their organisations. They tend to expect generic research to provide practical answers to their own problems directly. This is usually not the case. On the other hand FM practitioners and consultants frequently overstate the importance of their individual case specific results, assuming that the findings from one study for one organisation at one time, will have generic credibility and will be capable of broad application to others in the field. Academic researchers have tended to compound this confusion. Few have any direct practical experience of FM, so they risk disappointment when their ideas and concepts, the findings of their research and the generic knowledge produced, prove difficult to implement in the real world.

Figure 46 provides a basic framework through which the vital linkage between practical information and generic knowledge may be achieved along the Knowledge Trail. The Link A (Specific to Generic) shown in Figure 46, relates to those in advanced FM practice who will need to feed forward their specific experiences, critically comparing and developing the results of their work alongside that of others, in order to contribute to the FM knowledge base. Link B (Generic to Specific) in Figure 46, relates to those in the research community who will need to take responsibility for the interpretation, dissemination and exploitation of generic research findings, adjusting them to meet the specific circumstances and requirements of any particular organisation to facilitate implementation. Figure 46 shows the five main areas for the development of these links, all of which will be vital to the future success of initiatives to develop the FM Knowledge base:

Figure 46   Generic Knowledge and Situation Specific Information.

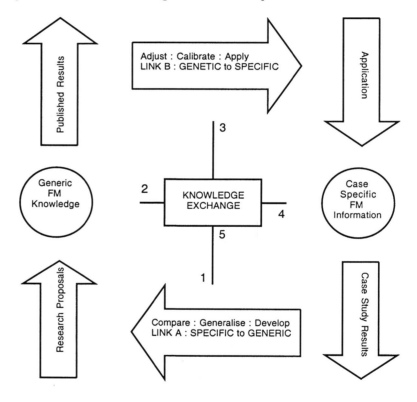

1. Cross-sector bench marking, documented case results, expert position statements and the identification of key operational issues for research.
2. Collaborative research with FM practice.
3. Feasibility studies, technical evaluations, software development and applications research.
4. Expert procedures and decision protocols to support FM functions and operations.
5. Knowledge Exchange.

A fuller account of these areas of opportunity has been published elsewhere (Nutt, 1999). Initiatives across the territory illustrated in Figure 46 will help to establish a consolidated FM knowledge base, supported by secure techniques, sound criteria and accountable expertise. The knowledge exchange link will be of central importance. It will provide the basic interface for the exchange of FM experience, expertise and data, between those in advanced FM practice and consultancy, the providers of FM services, those in FM education and training, and the emerging FM research community, focusing on key questions for the future:

- What needs to be done to build a secure international knowledge base with supporting techniques to meet the challenges of facility management futures?
- What could be the unique contribution of FM knowledge in the future to organisations and individuals, in terms of scope, skills and techniques?
- How can its potential market niches be identified?
- How might facility management practices be further impacted by 'electronically' based business environment?
- What forms of process and management knowledge might enhance service reliability and add value to FM operations?

Questions such as these are the focus of the next five chapters of this book. Chapter 22 considers the impact of communication and information technology on 'intelligent working'. It gives a synoptic account of the emerging knowledge-based forms of working and the new approaches to their management so that groups, teams and individuals can work independently of location, place and time, as introduced in the second part of this book. The range of intelligent work activities are discussed in some detail. Here, futures along the HR and Knowledge Trails promise to be highly compatible and reinforcing. Facility management and facility design face a similar challenge, the provision of fast response working environments and short duration business space and infrastructure, to provide high quality 'work-space' and 'work-time' settings to support the knowledge-based businesses of the future. Serviced office provision is an primitive example of the type of flexible support arrangements that these new business activities may require.

Two related themes run through the following chapters. First, how to create, and then realise, the value of the FM knowledge base, both for organisations and individuals, even while the content and structure of FM knowledge remains unclear, as at present. Second, how to develop appropriate and viable techniques for the delivery and implementation of this knowledge through facility management and facility design. Two generic models that are central to the development of the FM knowledge base, are explored next.

Risk evaluation and risk management have been one of the key defining decision concepts of the last twenty years. The issue of risk, both at a strategic level in relation to organisational competition, success or failure, and at an operational level concerning human health, safety, security and well being, has impacted on all levels of management. It now underpins virtually all FM functions, from basic procurement appraisals to formal risk assessments concerning use, compliance and management under national and international statutory instruments. Chapter 23 includes an account of the initiatives that are underway to develop a generic risk database for facility management.

The development of generic techniques is considered further in Chapter 24. It provides a summary account of FM measurement systems and benchmarking. During the formative years of FM, benchmarking has clearly been a useful tool. It is still of value for newcomers to a business sector or to a specific area of FM practice. But benchmarking can have limited value thereafter. What is 'best practice' for some can be 'inappropriate practice' or even 'poor practice' for others, given the specific circumstances of their organisation and business. Cross-sector benchmarking, on the other hand, promises to be of considerable value for the future. Chapter 24 provides a concise but comprehensive overview of cross-sector bench marking from an international comparative perspective. The concept of quality management underlies this discussion.

Chapter 25 explores performance measures from the perspective of facility design strategy. It describes the results of Australian research into the ways in which design knowledge, particularly workplace design knowledge, can contribute to work performance and morale, and to change management generally.

The final chapter to this part of the book describes the unique potential and strategic advantage of FM's intellectual capital, outlining the ways in which it could be exploited in the future. Facility management tends to serve two rather divergent sets of interests; those of the facility owner or investor and the needs of the facility user or customer. This places FM in a potentially powerful position. It should exploit its intellectual capital, realising the unique strategic value of the FM knowledge base, linking as it does, the knowledge of physical facility performance with the knowledge of business objectives, operations and support services. The generic knowledge chain that links these two divergent areas of concern is described, highlighting the critical importance of the business brief, the operational brief and the design brief in the future exploitation of the facility knowledge base.

References

Baladi P., (1999) Knowledge and competence management: Ericsson business consulting. *Business Strategy Review*, vol. 10, issue 4, pp. 20-28.

Edvinsson L. & Malone M. (1997) *Intellectual Capital,* Piaktus, UK.

Stewart, T (1997) *Intellectual Capital*, Nicholas Brealey Publishing Ltd, London.

Nutt B. (1999) "Linking FM Practice and Research", *Facilities*, Vol. 17, No. 1/2, Jan./Feb., MCB University Press, pp. 11-17.

22 Accommodating Future Business Intelligence

Stephen Bradley, Space for Business
Geoff Woodling, BFN Rebus

This Chapter examines how intelligent systems are reshaping the way organisations function, and their relationship with the environments they occupy, both individually and collectively. The Internet and the huge opportunities it is creating was virtually unknown outside the academic world until 1994. The consequent changes in demand for and management of business space are also yet to come.

In a volatile economic climate, most organisations are controlling costs very actively and are therefore seeking to minimise fixed costs of business space and information and communications technology (ICT) infrastructure. However they often need rapidly to accommodate fluctuations in number and location of personnel (often project-based) for a restricted duration and ideally without incremental increase in support costs.

In many organisations, large projects are expected to commence almost overnight and yet existing corporate infrastructure is already being run as 'lean' as possible leaving little capacity for unexpected headcount change (up or down); in many cases new ways of managing space and time is also well established and supported. In cases such as these, how can new infrastructure be obtained when and where it is needed without constituting an extra incremental fixed cost? How can alternative workplaces maintain the corporate values of identity, service and environmental quality as well as enabling rapidly assembled teams to form and perform effectively from the outset.

In traditional procurement of workplace accommodation there are a number of gaps between the generation of demand and the production of supply solutions. These are some:

- The property, construction, communications infrastructure and facilities services marketplaces are mostly separate.
- 'Procurement' of space and services is often an ad-hoc affair with lessons to be re-learnt and relationships to be established each time.
- The owners and managers of business space are often unresponsive to the speed and short duration of business demand.
- The design and fit-out marketplace is generally geared to deriving value from bespoke corporate solutions.
- The workplace furniture industry thrives on pushing new fashions rather than supporting integrated accommodation solutions.

Attacking the conservatism of the traditional markets in space and services, new operators such as Regus and HQ Business Centres are rapidly growing their brand and their portfolio of internationally consistent, short term serviced business space. Yet their offer comprises but a small percentage of the total business space of Western Europe, and their offer is only recently starting to provide infrastructure for project teams rather than itinerant sales forces or start-ups. There is scope for many other offers in the marketplace, with a perceived growing and unsatisfied demand for fast response, limited duration business space and infrastructure. The parallel demands by corporates for new types of limited duration serviced space alongside conventional, core space are analogous to rapid response, generally shared access, 'packet-switched' internet services, as opposed to the 'serial-switched', sequentially programmed response of the traditional marketplaces in network access. Who will be the players in the new marketplace for packet-switched space, and from where will they draw their inspiration?

> "Every few hundred years in Western history there occurs a sharp transformation. Within a few short decades, society rearranges itself—its world view; its basic values; its social and political structure; its arts; its key institutions. Fifty years later, there is a new world. And the people born then cannot even imagine the world in which their grandparents lived and into which their own parents were born. We are currently living through just such a transformation."
> (Peter Drucker, 1993)

New Information, Communication and Entertainment (ICE) technologies are shifting the boundaries and behaviour of markets. Their impact is encouraging the following phenomena:

* Change in the content and purpose of work from "labour" to 'knowledge leverage'.
* Interactive work with multiple people in multiple places, synchronous and asynchronous.
* Changed threshold scale at which people need to form formalised organisations.
* Undermining of control based on traditional hierarchies and information possession.
* Closer relationships between producer and consumer (the dis-intermediation of markets).
* Integration of information networks between customers and suppliers.
* The outsourcing of activities which are not unique to the value proposition of an enterprise.

These phenomena are not the drivers of change, but the implications. The consequence for organisations of these trends is that the 'information density' of work is increasing enormously and with it the demand for greater capacity to communicate and process information. The consequence for individuals of these work trends is certainly to permit many of us to work independently of location and time. The drivers of change are perennial: the search for defensible competitive advantage from applying innovations to stimulate new demand; new channels for exchange of goods and services; and control of resources, markets or environments.

Organisations compete to occupy places that confer the greatest advantage. The big difference from the past is that these 'places' no longer have to exist in real, fixed locations, we only have to 'rent' access to facilities and resources wherever they may be, whenever needed. The ability to gain access 'anytime, anywhere', is a fundamental aspect of new ways of working, and a major attribute of the quality of place (the genius loci) in the new economy, in which speed of access will be the major provider of comparative advantage.

Demands for the manipulation of complex information cannot be accomplished solely from a 'personal communicator' device while roaming from place to place. The broad-band infrastructure required to support specialist professional workers (ranging from doctors to designers) will need to be made available on demand in diverse workspaces, much as research facilities are made available to scientists today. The new 'information lab' will combine multi-media data interfaces with access to proprietary on-line software plus technical, environmental and social support for users. It will not be the exclusive preserve of software engineers but will provide the engine room of a diverse knowledge economy whose products will entertain, inform and manage our lives.

Workplace Procurement

New forms of work process and new forms of employment, in new types of organisation and whole new industries, are generating new demands to accommodate business activity alongside learning, residential and social activity. Released by ICT from the traditional constraints of work-time and work-place, new 'communities of interest' are starting to demand a richer and more dynamic balance of work (paid and unpaid) with learning, leisure and domestic life. This is a revolution in the sense of a return to a view of work better integrated within the wider context of society and life. Where, how and in what kind of place the new forms of working, learning and living will be accommodated, is now an open question. There is a need to re-think the meaning of 'office' space and look for a closer integration of land use and transportation planning with commercial, leisure and residential development policies.

The office as an administrative 'factory' will become a redundant concept – a controlled environment for sheltering Taylorist processes, best exemplified by the current 'year 2000' model of the densely populated, efficiency-driven call centre. The future workplace will be replaced by a broad diversity of places for specific activities:

- Network access, (physical and virtual).
- Meetings, conferences.
- Entertaining, recruiting, bonding.
- Acting, playing, forecasting, tracking, learning.
- Diagnosing, designing, modelling, testing, viewing, comparing.

- Fabricating, assembling.
- Guarding, defending, negotiating.
- Trading, exchanging (ideas, goods, securities, options, cash flows).

However it seems apparent that the majority of corporations have yet to provide appropriate support for increasingly diverse work practices. The opportunity of ICT-enabled new ways of working is the creation of a multi-dimensional system in which the worker is able to achieve more, individually and collaboratively, than is possible when tied to a singular workplace. To achieve this there has to be a 'win-win' sharing of responsibility and benefits between individual and organisation.

The high performance workplace as a 'club' for highly autonomous and highly interactive collaborators, as described by John Worthington and others in "Reinventing the Workplace" (1997), and by Frank Duffy and others in "The Responsible Workplace" (Duffy et al., 1993) and "The new office" (Duffy, 1997), is of course still rare. The development of mixed-media environments: computer-supported-collaborative team workplaces for colleagues who are often physically remote but congregate to learn from each other from time to time, are rarer still. In 2000, such tele-presence hybrids of the virtual and the physical are as yet confined to media-labs in technology and communications companies, high-end management consultancies and research institutes.

Intelligent Communities

The new choices of place available to organisations as well as to individual workers is a matter of increasing importance to business and political leaders. Information and communication technology (ICT) enables new choices for both the individual worker and the organisation. The new ways of working slogan 'anytime-anyplace' assigns much greater importance to the quality of place than hitherto in terms of support for work style and lifestyle choices, both individual and collective. Depending on the characteristics of their workforce and customer interaction, some business activities will thrive in regenerated urban centres; others in suburban or rural environments, both enabling the re-integration of work style and lifestyle.

The comparative advantage of 'intelligent communities' will be influenced greatly by the quality of available communications infrastructure, in particular the density of high-bandwidth cellular radio to support increasingly information-intensive mobile business connections (including video interactivity). It will also be determined by availability of a rich hierarchy of physical work settings, offering user choice of technical, ergonomic, social and psychological support to the task in hand.

Future employment generation is fundamentally linked to residential supply as well as to critical mass of competing enterprises, but is also dependent on other community factors. Educational, health and leisure amenities, political, social and environmental characteristics and 'connectivity', both physical and electronic, are major determinants of comparative advantage for urban communities.

As examples of new 'clusters' or communities of interest, the new multimedia communities south of Market Street in San Francisco ('Multimedia Gulch') or in London's Soho and Clerkenwell are literally 'pulsating' as new creative businesses form and re-form as new breakthroughs are invented and exploited. The participants congregate where they find the social and environmental stimulus and quality as well as the partners or suppliers, and where they can connect with customers and financial capital markets. Intelligent communities which have thrived and became famous, in a virtuous, self sustaining cycle have always had the following characteristics:

- Well-located, with attractive and comfortable climate (political, social, aesthetic and meteorological).
- Consistently well-managed, and well-resourced.
- Able to connect together a critical volume of willing buyers and sellers.
- Smart enough to anticipate and adapt when the basis of exchange was substituted by a new commodity or technology.

The rate of information exchange will become the defining quality of wealth creation, wherever people live, work and play. The greater the rate of information exchange, the higher will be the demand for all kinds of meetings. It matters little whether these are face to face in the same place or electronically linking people in different places. People will meet more, not less. Consequently, information infrastructure quality will emerge as complementary to the social and environmental quality of communities. Truly 'smart communities' will be where these qualities are conjoined with active management and with a clearly communicated vision of the relative position of that community.

Workplace Strategies In "Reinventing the Workplace", by Becker and Joroff (1993) the workplace is seen as a system:

> "In essence, the IWS concept characterises the workplace as a unified system. It is a system that creatively combines wisdom about the nature of physical settings (where work is conducted); the information technologies used in the performance of work (how data, opinions, and ideas are accessed, processed, and communicated); the nature of work patterns and processes (when and how tasks must be performed to achieve business objectives); and finally, organisational culture and management (the formal and informal values, expectations, policies, and behaviour that influence all of the other factors)".

Many corporations have found their ability to change compromised by holding and managing their own private workplace infrastructure. Increasingly project-driven, fragmented organisations need to be able to procure their accommodation and network access on demand, and more autonomous end-user managers in the operating units are requiring to take facilities decisions closer to the local markets. There is an increasing tension between the changing nature of organisational demand (to accommodate a fluctuating size and location of operational workforce) and the ability to obtain appropriate solutions from the market. The future may lie in facilitating the market for business accommodation in the organisation by informing, not controlling their clients' options.

Viewed in this light, the static concept of place does not fit well within the organisational framework of the agile, de-centralised enterprise. Although the 'integrated workplace' concept is viewed as a system of settings, infrastructures, processes and behaviours, it is defined within the context of a unitary occupying corporation. As we have argued, the organisation as we know it will change in ways that give individuals and small groups the opportunity to redefine their roles, relationships and identities at different times, to reduce the distinction between social and work roles, or even of being 'in or out of work'. Work and employment is no longer defined by belonging to a single organisation.

It is open to question whether the de-centralised, networked corporation will continue to be able to pursue corporate infrastructure strategies to allocate space and facilities to meet the diverse demands of the many and fluctuating organisational elements beyond a small core of strategists and contract administrators. The networked organisation can respond to changing and transient needs faster and more efficiently as a self-organising system. What purpose would a corporate infrastructure strategy serve other than to distort the process of adjustment or of course to preserve control? This is already the challenge for corporate real estate managers who find their internal clients reluctant to accept the impositions of corporate guidelines.

The Attractive Workplace

People now may 'belong' to several networks or teams, connecting with different organisations in different places at different times. Space and place must cater for changing behaviour: who we want to be, what roles we want to play and with whom do we want to be at a particular time. These are the attributes of new behaviours that organisations must learn to support.

Released from traditional constraints of work-time and work-place, different segments of that fortunate, mobile population which is able to exercise choice will choose to occupy multiple places in well serviced, lively inner city areas, suburbs and rural environments. Their choices will be driven by environmental and social motivations rather than solely by employment opportunities, by opportunity rather than simply by need. Better understanding of these segments and their patterns of space usage for diverse, yet often overlapping activities, is now needed equally by community planners, by business leaders and by the property industry.

New types of business activity, in new 'communities of interest', tend to re-colonise older, cheaper and often more adaptable inner-city property, both for working and living. Transportation constraints, environmental and leisure preferences will predominate. 'Connectivity' and density of information flow, both physical and electronic, will be major determinants of value and demand. Accessibility and quality of public spaces and leisure, entertainment and hospitality amenities will provide major comparative advantage for city districts. The economics of supply (place) and demand (fashion) will ensure that not everyone will be living in ski resorts or in prime inner-city riverside apartments.

The possession of choice will be the new status symbol, the new wealth. Those able to exercise choice will be able to occupy more not less space, more intensively, in smaller increments of time: truly 'packet-switched' rather than 'line-switched' space, analogous to the command of bandwidth on communications networks. Elitist this may sound, but those able to exercise choice are the generators and exploiters of new ideas, new technologies and new forms of economic value generation

Cloning Communities

Competition among communities is the process at the heart of evolution. What makes communities attractive and competitive? Can the competitive success of business communities be cloned in different locations? A community that has developed high capabilities of both infrastructure and human resources, and also has both a high level of understanding of its comparative advantages, together with a purposeful leadership, is an 'intelligent community'. What are the qualities of communal intelligence that can be cloned?

The operating environments for such intelligent communities or 'information-rich' work environments will be multi-layered. At the global level it is that of world regions which attract and retain high value generating industries; developers, producers, suppliers, manufacturers, assemblers and distributors of information products, entertainment hardware and software). At the local, micro level, it is that of the 'connection space' in main street or transport interchange; the function which used to be fulfilled by the hostelries, taverns and coffee-houses of the market or souk.

The penetration deep into the organisation of market forces in space and facilities services will challenge many existing monopolies of space 'ownership' and the way that organisations occupy space. The changing power balance between fast-adapting, smaller, 'agile' companies (or high agility, subsidiary parts of larger enterprises) and large institutions will have inescapable impact on property markets.

The built environment within existing communities will have to evolve to provide new forms of workplace. The delivery of property services in 'intelligent communities' and the means of financing services and physical assets, will evolve to provide more liquidity to new enterprise cultures. A market is already evident for those who already control the resources of real estate, financial capital and network infrastructure. To these resources need to be applied the leverage of management, design, marketing, logistics and customer service skills.

As the mobile workforce, either as individuals or in teams, demands more and more access to communications bandwidth and specialist software, the need will grow for 'information labs' as limited duration work facilities, providing comfort, conviviality, technical support and possibly lifestyle services for the 'time-deprived'. Short-life cycle business enterprises would choose not to develop, own and occupy such high cost network assets on their own, unless they can gain competitive advantage by depriving competitors of such resources. Third party providers will also undertake the development, financing and management of such specialist space incorporating high bandwidth computing and communications facilities offering time-shared access for the new generation networked enterprises, who will thereby be freed from the barriers of having to finance such assets.

Commercial property ownership and development will become merely a component of a new business sector. The short-stay, flexible offer, high value-added serviced accommodation market is as yet only 5% of the total in the western world yet is predicted to quickly reach 15%. As new accommodation providers bundle ICT infrastructure together with space and services, a new generation of enterprises will be freed from the limitations of time and place imposed by traditional real estate and technology property commitments.

Who will be the market-makers in the 'networkplace' of the future?

References

Becker F. & Joroff M. (1993) *Reinventing the Workplace*, Corporate Real Estate 2000, IDRC, USA.

Drucker P. F. (1993). *Post-capitalist society*. Harper Business, New York.

Duffy F. et al. (1993) *The Responsible Workplace: The Redesign of Work and Offices*, Butterworth Architecture, London.

Duffy, F & Powell. K (1997) *The new office*, Conran Octopus, London.

Worthington J. (Ed.) (1997) *Reinventing the Workplace*, Butterworth-Heinemann, London.

23 Developing a Generic Risk Database for FM

Barry Holt, Willis Group
Andrew Edkins and Germán Millán,
University College London

In an increasingly competitive business environment all organisations are continually faced with a wide spectrum of risks and uncertainties which can affect the achievement of their objectives. The distinction between risk and uncertainty is important, and depends on the level of information that is available at the time when possible outcomes are being considered (Galbraith, 1977). Risk is normally considered to be present in those situations where sufficient data exist and enough information is available to be able to assess the factors involved and therefore to develop some form of quantification of their effects. These quantifiable risks are of a very varied nature and require the response of a diverse range of strategies.

In broad terms, the risks that business organisations and FM organisations in particular, face can be classified in two groups. The first kind of risks may be defined as pure risks. They are traditionally thought of as downside loaded, implying threats to the survival of the business itself. The only possible outcome here is a loss to the business or a failure to achieve its objectives. These risks include the threats with which all are familiar, such as those which could have negative impact on the physical assets of an enterprise or those that could cause harm to the employees, the public or the environment. An illustration of pure risks can be found in a recent incident at London Underground. A driver allowed a tube train to roll backwards for a considerable distance before fail-safe systems halted it at a station. Such an episode, in this case thankfully with no injuries, can only have negative consequences. Those in charge will no doubt be required to ensure their systems and procedures will be set up to eliminate the possibility of another similar accident.

The second broad kind of risks can be defined as speculative risks. In this group the outcome could have either a positive or a negative impact. In fact, they can be understood as a form of gambling. Whilst many would relate this to money won or lost on games of chance, there is a great sophistication in the 'academic' understanding of such speculative risks. This can be found both in business and economics research and in particular, in areas such as game theory. Classical examples can be found in association with foreign exchange rate fluctuations or the effect of company take-overs. An interesting recent example of a speculative risk was the tender for the licences to operate the next generation of mobile phones in the UK, which consisted of a multi-stage auction devised by game theorists. The outcome of the auction was a multi-billion pound windfall to the UK Exchequer. To know whether the price paid to hold such a licence will allow the successful bidders to make a profit necessary to keep them in business in the long-term (with a contented set of equity owners), the successful companies will have to wait several years.

Different types of decision agent tend to have different perceptions of risk and adopt different approaches for its management. For example those responsible for running business-based organisations recognise that all risks, both 'pure' and 'speculative', need to be dealt with effectively within a risk management process. Risk management in the area of finance is rather different. It usually relies strongly on the concept of a portfolio. The larger the set or portfolio of investments belonging to the decision maker, the more likely it is that the actual outcome of their portfolio will approach its expected value. This means that one of their central aims is to diminish the variability of the portfolio, while increasing its value overall. In contrast to the 'business' and 'finance' viewpoints, a large proportion of the literature in project risk management and many practitioners of risk management, assume a definition of risk that is consistent with the definition of pure risk. In the construction industry generally, and in the field of facility management in particular, agents tend to perceive risks in a rather different way than those in finance. They seem to perceive speculative risks as if they were pure risks. In other words, risk managers perceive the speculative risks as if they were downside loaded.

It is therefore essential to ensure that the managers of project-related risk should focus both on the downside and the upside risk. The former is associated to what Chapman & Ward (1997) call "a threat to success", and the latter to 'opportunities' arising from the project. The same authors recognise that it seems a natural step to regard risk management as essentially about removing or reducing the possibility of under-performance. This concentration on avoiding negative outcomes is compounded by a failure to consider the possibility of better-than-expected outcomes.

Traditionally risk management has followed a structured approach with a number of distinctive stages. In fact, there are several risk management models but they all tend to be built around a three-stage process, such as the one proposed by Brandon (1990). The traditional or normative approach to risk management starts with a phase of risk identification or hazard identification. Typically, at this stage, sources, events, effects, types, likelihood and severity of risks are identified. Second, it is considered a stage of risk analysis, when a quantitative assessment of the previously identified risks is carried out. And finally, a phase of risk response or risk control is considered. In other words, once risks are identified and evaluated, decisions are taken as to what to do about them.

Typically, strategies for reduction, avoidance, transference or retention are decided; namely, 'eliminate it'/'ignore it', 'transfer it'/'insure against it' or 'absorb it'. It is clear that the decision about the risks needs to be communicated to all of those who may be involved in the risk environment. This process is summarised in Figure 47.

It is acknowledged that there are many variants of this model, some of which consider larger number of stages, an example being Chapman & Ward's nine-stage model. They are however, variations on a common theme, rather than fundamentally different.

A number of approaches of the kind shown in Fig. 47 have been developed to respond to the common hazard based threats. These approaches present valid ways of looking at the technique by which we can manage the various form of threats. However, if a broader view of risks is considered, and their impact on the business as a whole is the main concern, then, this approach has a number of disadvantages. These arise both in terms of effectiveness and of ease of application. In particular, within the area of facility management, these disadvantages can lead to inefficient or inappropriate risk management. This may happen because many of the threats arise not only from physical causes, but also from the way business relationships are developed.

Figure 47   The Risk Management Process.

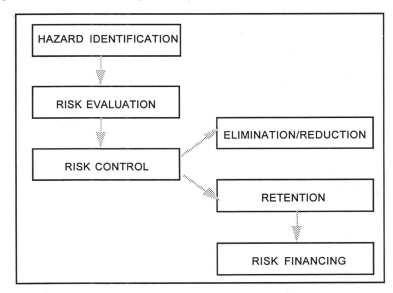

In summary, traditional approaches rely on the identification of threats in isolation. They then attempt to evaluate their potential impact and estimate the likelihood that these threats will generate a loss. This approach has a number of disadvantages, which can result in risk management being seen as a negative factor in the business. In particular, the traditional approach:

- Is frequently not aligned with the business objectives of the enterprise;
- Is carried out as an 'add-on' exercise rather than as part of general management;
- Addresses the threats at a stage where it is often too late to take the most effective action.

This chapter discusses an alternative approach to these risk management techniques. A more effective approach might be to begin with the objectives of the business, its operations and activities, to establish the factors that must be in place for these to be achieved. If adopted within the facility management field, this alternative approach could enable businesses to achieve their goals more efficiently through increased knowledge of the source and consequences of all risks that may arise.

Techniques

A wide range of techniques can be used in practice to carry out the identification and evaluation of threats and risks at each of the above stages. However, the greatest drawback with most of these techniques is the absence of a concise methodology for assessing the project delivery process. The three techniques that could offer the greatest benefit are proposed to be:

- Brainstorming and the risk register;
- Key risk profiling and
- Dependency modelling.

Each of these techniques will be outlined in turn, with a summary discussion of the potential advantages and limitations. Following this, the use of a 'risk database' is described and analysed within the context of the three risk management techniques. Finally the potential value of these tools within the field of FM is discussed with some conclusions for the future.

Brainstorming and the Risk Register

Brainstorming and the production of a *risk register* constitute a traditional generic approach to risk identification. In fact, Chapman & Ward (1997) regard brainstorming as one of the best and well known techniques used for fostering creativity. Through the provision of more possible solutions and unusual approaches to a problem, the problem analysis is improved. This is often undertaken at the planning stage and it involves selected individuals for random identification of threats. The output would be in the form of a risk register, which may be prioritised.

This falls short of a robust, effective and efficient methodology for identifying and managing risks in projects. However, the greatest drawback is that detailed threats cannot be wholly determined at the front end of the project. Moreover, the interdependencies of these threats within the critical project processes cannot be fully determined. The principal advantages and disadvantages of this technique are shown in Table 14.

Key Risk Profiling

Key risk profiling (KRP) is a form of facilitated workshop undertaken with the developer, which identifies the threats to the business objectives. This could be applied at the stage of establishing the needs, or otherwise, of project planning. The output from KRP would be a prioritised list of the key threats and identification of the source from which they could arise. It has the advantage of being much more refined when compared to the brainstorming and risk register approach. However, its main drawback is its failure to generate a risk map that identifies and reflects the critical project delivery processes. Hence, it neglects the importance of the interdependencies of threats to the project. Its principal advantages and disadvantages are shown in Table 15.

Table 14   Advantages and Limitations of Brainstorming.

Advantages	Limitations
Simplicity and ease of use	Developer and other key parties on the project are often excluded from the process
Quick	Threats not always indentified at the front end of the project
May not require specialist knowledge	Does not consider interdependencies of threats and hence critical project processes
Involves more than one individua	Limited quantification
	Does not identify resources and responsible personnel for management of the threats
	It often does not fully reflect the business objectives and expections

Table 15   Advantages and Limitations of Key Risk Profiling.

Advantages	Limitations
Developer/end user is involved in the process	Time consuming
Identifies the sources from which the threat can arise	Not quantitative
Uses a multi-disciplined team	Needs specialist facilitators
Considers business objective	Does not consider interdependencies of threats and hence critical project processes
Identifies existing and required mitigation	Cannot consider the sensitivity of the process to mitigation

Dependency Modelling

Dependency Modelling is a structured methodology that identifies the critical resources and processes on which the successful delivery of the project depends. The output from dependency modelling is in the form of a logic tree (risk map) which displays the interdependency of events. This technique outlines the critical processes required for the successful delivery of the project. The resultant logic tree (with identified threats) can be used to identify the project failure exposures and the concentration of threats. It can also identify the likely failure pattern and worst combination of failure patterns. Its main advantages and disadvantages are shown in Table 16 and an example of a risk dependency map is illustrated in Figure 48.

Table 16   Advantages and Limitations of Dependency Modelling

Advantages	Limitations
Done at the front end of the project	Requires specialist facilitator
Identifies key processes from which risk can arise	Time consuming
Considers interdependencies of threats and processes	Some quantitative input required which could be subjective
Considers business objective	
Involves a multi-disciplined team	
Identifies existing and required mitigation	
Identifies the sensitivity of the process to risk	
Considers effectiveness of the combination of controls	
Identifies critical failure mode	
Calculates the business exposure resulting from threats within the business processes	

Figure 48　A sample risk dependency map.

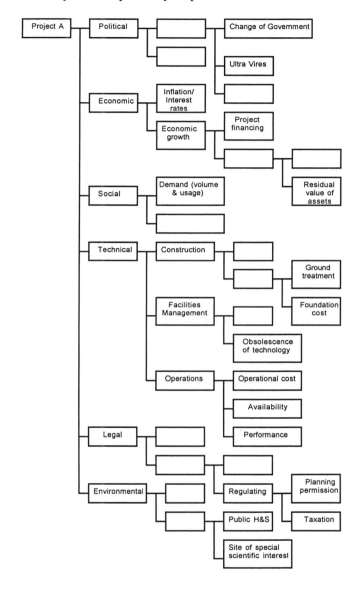

This risk dependency map illustrates not only factors which contribute to the success of a project. It also identifies those that can be seen as a potential threat, allowing the interdependence of these success/failure factors to be established. For example, in the map in Figure 48, a facility management issue is the obsolescence of technology. In different circumstances this could also have an impact on operational and environmental exposures. The ability to identify interdependencies is of significant value in assessing the likelihood of a project's success or the consequences of its failure. The traditional approach forces the focus on the individual threats. Nonetheless, the greater part of major losses have arisen through the interaction of several factors, which being inter-related have not been identified. Another powerful benefit of establishing such a risk map and identifying the relations between factors lies in the ability that it gives us to evaluate the likelihood of the particular scenarios.

In many cases, traditional risk assessment has lacked a rigorous foundation. The threat is often classified simply as 'high', 'medium' or 'low' risk on the basis of what is essentially a subjective judgement. Again, this indicates the boundary between risk and uncertainty. When looking at the management of major projects, this is an extremely uncertain basis on which to base decisions. In other areas of both project and general management such decisions would require far more detailed analysis.

The next logical development is an event tree showing the manner in which events must combine for the successful project outcome to be achieved. The events are related by 'and' or 'or' statements in the same way as in fault tree analysis. If it is possible to establish the probabilities of success or failure for the individual components, the probability of the outcome can be established very simply. This will take into account the interdependencies that have been identified in the modelling phase. In addition to considering each activity and each new threat for each project it has been possible to use the data from a large number of risk assessments to establish a database of threats and their potential impacts.

Risk Database

As a result of a large number of risk assessment projects which have been carried out across a wide range of industries, Richard Oliver, now part of Willis Group, have been able to develop a risk database which provides a powerful tool to aid the evaluation of risks to any business activity. This database is based on a commercially available software program, KnowRiskPro® (CorPorfIT), which contains two principal functions:

(i) A knowledge base, which contains data concerning:
   • The risks facing the organisation;
   • The impacts or consequences of those risks and
   • Controls on those risks or on their consequences.

(ii) A series of three types of risk profile, namely:
   • A business unit profile for tracking the business units' exposure and progress towards achieving risk mitigation targets;
   • Template profiles, which link common risks, consequences and controls for inclusion in a business unit profile and
   • Working version profiles for modelling 'what if' scenarios, which may form the basis of a template or business unit profile.

In Table 17 a sample from a report that was generated from the risk database is reproduced. Using this example to illustrate the discussion, the following points can describe the key features of the output of the technique.

Table 17 A sample of report generated from risk database.

RISK	POTENTIAL IMPACTS	RISK ASSESSMENT (INHERENT)		INHERENT MATRIX RATING	CONTROLS	RISK ASSESSMENT (RESIDUAL)		RESIDUAL MATRIX RATING
		F	S			F	S	
Strength of sterling	Loss of customers Loss of direct exports Loss of PBIT Loss of sales in home marke OEM exports Uncompetitive	5	4	Very High	Developing markets where currency exchange more favourable Forward purchasing - Group policy Rebate customers - exchange rate limited (specific customers) Survey of global markets especially Asia	4	3	High
Damage via fire, vandalism to key processes	Emergency replacement cost Failure to deliver products on time and loss of customer base Lost PBIT Lost sales Unable to supply customers product	3	4	High	Controlled key holder/ access Disaster recovery plan Fire precautions - drills/extinguishers/trained fire marshalls Health & Safety procedures/risk management Physical security procedures/alarms Specialist security on IT facilities	3	4	High

(a) Risk

This relates to the threat to the business activity that has been identified through one of the techniques described above. The assessment is likely to generate numerous risks in this column, the example shown representing only a small portion of the final output. All risks are incorporated, not only the traditional 'hazard-based' threats such as fire and accident; but also the risks associated with the business, which may be either positive or negative risks. The example illustrated in Table 4 is the strength of sterling, which depending on the direction it moves, may be either a threat or a business opportunity. In this case there were potential downside risks as sterling was felt to be too high.

(b) Potential Impacts

Given that the risks facing the organisation have been identified, the database will generate the range of potential impacts on the operation. For example, the strength of sterling could obviously have a significant effect on direct exports. However, in the home market there could also be a loss of market to foreign imports, which would have become cheaper. Similarly at the operational level, the effects of a fire or vandalism to key processes would not only entail replacement of the damaged equipment, but also it could result in long term loss of customers through the inability to supply or to supply on time.

(c) Risk Assessment (Inherent)

From the evaluation of risk, the individual threat is assigned a rating (1-5) in terms of likelihood, foreseeability (F) and severity (S). These generate a rating through the use of a two dimensional matrix.

(d)   Inherent Matrix Rating

At this stage the inherent risk, is considered i.e. the level of risk, assuming that no control measures are in place. The matrix presents a qualitative rating, which can be used for assessing the potential viability of a project or business activity.

(e)   Controls

The database will generate a list of possible controls, which can be applied and will allow the original inherent risk rating to be revised to give a residual matrix rating, which represents the degree of risk remaining with controls in place.

At the present time Willis' database contains around 15,000 identified threats, their potential impacts and controls. These have been gathered from a wide range of industries and, if used in their basic form for an activity such as facility management, they would lead to unmanageable quantities of data. However, the database management software allows for filtering of data to produce industry profiles. The creation of a facility management risk database would simply require the creation and application of an appropriate filter and the setting up of a suitable business profile. At a practical level, Willis' database is continually being refined and updated. In particular, the possibility of applying statistical techniques to test the validity of the information and to identify correlations between different risk areas is currently being investigated.

Future Directions

The generation of a risk database does not diminish the importance of an individual's perceptions about the risks or hazards. Indeed, the development of a more rigorous and data-rich system should act as an aid, rather that replacing individual and group considerations.

Recognising this an on going research project, at University College London, is exploring the way in which those with responsibility for construction projects actually perceive the risks that are present in their projects. To achieve this it is necessary to look at the cognitive filters and biases that the individual experiences in understanding and responding to the world around them. Furthermore, it is necessary to examine the influences that groups have on these perceptions. The research, using expert cognitive based methods, aims to understand how risks are perceived and, if possible, to test for the individuals' appetite for risk within construction projects.

The results from this research will bolster the way in which the risk identification process is understood. The incorporation of the results within the techniques outlined in the previous sections, will hopefully improve their reliability and establish a secure basis for the next generation of generic risk data bases for FM.

Concluding Comments

The worst type of risk that an organisation can face is the risk that has not been anticipated, identified or evaluated. Decisions taken without due regard to the wide range of threats that may be encountered face the probability of major loss even the possibility of failure. The main obstacle to carrying out adequate risk assessments for all projects is usually the time that this would take. This is especially relevant when the nature of the projects, as managed by facility management organisations, are measured both on time and cost.

The initiatives discussed in this chapter, when fully developed, will result in a rigorous and more efficient identification and evaluation process across the whole range of facility management functions. They could also result in projects being managed in a way that minimises the possibility of objectives not being met, while at the same time, keeping the cost of risk management, in terms of both time and effort, to an affordable and practical level.

References

Brandon P. (Ed.) (1990) *Quantity surveying techniques*, BSP Professional, London.

Dale B. (1990) *Managing quality*, 2nd. Ed., Prentice Hall, Hemel Hempstead.

Chapman & Ward (1997) *Project risk management: processes*, techniques and insights, John Wiley & Sons, London.

Dickson G. (1995) *Corporate risk management*, Institute of Risk Management, London.

Flanagan R. & Norman G. (1993) *Risk management and construction*, Blackwell Scientific, Oxford.

Galbraith J. (1977) *Organization Design*, Addison-Welsey, Reading.

Godrey P. (1995) *The control of risk: a guide to systematic management of risk from construction*, CIRIA, London.

Kumamoto H. & Henley E. (1996) *Probabilistic risk assessment and management for engineers and scientists*, IEEE Press, New York.

24 Measurement Systems: establishing cross-sector validity

Tim Broyd and Roderick Rennison, WS Atkins plc

This chapter explores measurement techniques that can be used to drive cross sector construction and facility management innovation, world-wide. The construction industry is considered as well as the facilities management sector, as the two are inextricably linked. In particular, the impact of decisions made during the project planning, design and construction phases invariably have a major impact upon the whole life operation of the facility. The chapter focuses upon the application of benchmarking as a process measurement and improvement tool within the UK and US construction and facilities management sectors. It discusses benchmarking as a business tool, and its use within these sectors in the light of changing business and economic drivers.

In recent years, the term 'benchmarking' has become so widely used, that it has come to mean many different things to many different people. The definition used by Robert Camp, the author of one of the definitive benchmarking texts, is widely accepted, and is the one adopted here:

> "The search for industry best practices that lead to superior performance". (Camp, 1989)

The emphasis is upon comparing business practices, not just generating performance measures or 'metrics', though these are an integral part of any benchmarking effort. Thus benchmarking goes beyond simply establishing comparative measures without linkage to causes or processes. A further important point is that benchmarking is a continuous process, not a one-time effort. This definition suggests that benchmarking always involves looking outside one's own organisation to benchmark. However, this is not necessarily the case. There are many potential benefits from benchmarking internal processes, and in fact, internal benchmarking may be a pre-requisite before carrying out benchmarking with an external partner. In essence, benchmarking has become one of the key tools available to effect organisational change and drive continuous improvement.

Those companies which embrace benchmarking, believe that the rewards repay the effort and cost involved. Productivity, quality, and profitability are all improved. Rank Xerox is often considered to be the founder of benchmarking. In the late 1970s, it faced collapsing market share as aggressive competitors moved in and beat it on price, quality, and other key performance measures. Xerox used benchmarking to analyse the way its photocopiers were built. Wherever it found that someone else did it better, it adopted that competitors level of performance as a base for its own operations. As a result, it stabilised its market share, and saw a huge increase in customer satisfaction. Whilst the Xerox experience is well documented, in reality, companies have been attempting to measure and improve their business processes for nearly a century. Henry Ford, in his quest to build cars more efficiently, was probably one of the earliest proponents of benchmarking. Although some of the most high profile instances of benchmarking resulting in dramatic business improvement come from the manufacturing sector, increasingly construction and facilities management companies are starting to adopt benchmarking systems and seeing similar rewards.

The Benchmarking Process

The benchmarking process generally consists of five simple phases. These are:

- The Planning Phase - during this phase, the organisation identifies what is to be benchmarked, identifies comparative companies or business units, determines the data collection method, and collects the data.
- The Analysis Phase - the performance gap is determined, and future performance levels are projected.
- The Integration Phase - the findings of the benchmarking are communicated with a view to gaining acceptance within the organisation, and functional goals are established.
- The Action Phase - action plans are developed, implemented and monitored, and the benchmarks re calibrated.
- Maturity - the practices are fully integrated into the processes, and improved performance or superiority is attained.

Conceptually the benchmarking process is simple. This is not to say that it is easy. The actual process is complex, and to carry out meaningful benchmarking requires a significant investment of time and money.

Benefits of Benchmarking

Many companies believe that there are significant benefits from introducing benchmarking systems, including:

- Reduced overheads and increased profitability – many benchmarking exercises focus upon improving internal systems and processes, from administration and accounts, to recruitment and marketing. It is not uncommon to achieve a 20% cost saving in these areas.
- Improved Quality of Service – in almost every benchmarking exercise, quality and uniformity of service/product improves. Typically defects and rework reduce, and customer satisfaction increases (impacting repeat work performance).
- Improved Delivery Time – time savings of the order of 20 to 30% are not uncommon.
- Increased Staff Satisfaction – this impacts staff turnover and bottom line performance.

- Increased Technology Transfer – often, IT plays a significant role in technological development and transfer. Sharing best practices speeds the integration of new technology and technical expertise.

For example, in 1991, Skanska introduced an initiative called 'Think Total Time' (3T), which involved analysing business processes and implementing a benchmarking programme supported by training of its 15,000 employees. The results are impressive – average programme times reduced by 20% in two years, time required to resolve defects reduced by 75%, and 1 in 5 schemes signed off as having zero defects on final inspection. Other companies such as Tarmac ('Best in Class', 1998), Balfour Beatty (closely involved with the AGILE initiative) and Morrison (balanced scorecard approach based upon European Business Quality Model) have also introduced benchmarking systems into their businesses. For many companies an interest in benchmarking is triggered by a crisis within the company – be it reduced profits or falling market share. Leading companies talk of being one in ten by 2010, reflecting the view that there will only be ten global industry giants by that year. Benchmarking is one of the tools available to help survive in an increasingly competitive market.

Limitations of Benchmarking

Recently, benchmarking has been held up as the panacea for the ills of many industries, and the solution to every business's problems. It is not, though it is a powerful business tool. Benchmarking requires senior management commitment over an extended period of time, as well as an acceptance that current business practices may not be the best solution to every problem, even though historically they may have been successful. Carrying out benchmarking is likely to prove costly, given that it is a continuous process and requires significant resource allocation, and it may be difficult to quantify the financial benefits, especially in the short term. There is no guarantee of success; businesses may be so diverse that establishing any generic best business practices proves very difficult, or may even find that they do not appear to have any clearly definable best practices within the organisation. Additionally, even assuming that best practices and benchmarks can be defined, adopting these throughout a company may require a slow and painful cultural change.

Types of Benchmarking

There are a number of different types of benchmarking, some of which are likely to prove simpler to implement than others. The first type is 'internal benchmarking'. This is where the company focuses purely upon its internal business practices. The second is 'competitive benchmarking', where the company compares its business practices with those of one or more competitors in the same industry. The third type of benchmarking is 'industry leader benchmarking', where the business practices of the industry leader in the same industry are used as the basis for comparison. Finally, 'generic benchmarking' is where the business processes of an organisation in an unrelated industry are compared.

In most large firms, similar functions are carried out within different operating units, hence there is scope for comparison, or internal benchmarking.

This may prove a difficult task in very diverse organisations. However, if a company does not do this, it is not in a position to progress to any of the other types of benchmarking, because it has failed to identify its own best business practices.

The problem with carrying out internal benchmarking is that although it may identify best business practices within the company, there is no way of knowing how these compare with direct competitors or with other industries. Does the company in fact have the best practice? Therefore, once a company has carried out internal benchmarking, it should also consider benchmarking either against a direct competitor, or a world class industry leader.

Benchmarking against a direct competitor presents three fundamental problems. Firstly, one has to identify a comparable competitor. Secondly, there may be no way of knowing if they have best-in-class business practices. Thirdly, the confidentiality problems which might arise from sharing information may be insurmountable. Benchmarking with an industry leader may be similarly problematic.

Benchmarking with a world class company in an unrelated industry often brings about the most dramatic performance improvements, as a different industry may adopt an entirely different approach in solving problems or tackling certain processes. However, identifying a suitable partner who would be keen to benchmark with your company may be a problem. Recognised industry leaders such as Toyota and Xerox are inundated with requests from companies wishing to benchmark with them. However, they are naturally selective about whom they wish to share information with, and would expect to see a serious commitment to continuous improvement by their prospective partner, as evidenced by the use of internal benchmarking and other techniques by the partner.

Business Drivers

There is a growing feeling that for those in the construction and facilities management industry, the question is no longer 'Can I afford to do benchmarking?', but rather, 'Can I afford not to do benchmarking?' Clients are demanding improved performance, both in the UK and overseas. In the UK, many of the largest clients in these sectors are seeking dramatic improvements. Various forums have been established over the years in the UK, including the Construction Task Force, the Construction Round Table, and a variety of others. In the US, the Government is driving improvement through the National Partnership for Reinventing Government, formerly the National Performance Review (URL, a).

The Latham and Egan Reports

Both the Latham and Egan reports are very relevant to the facilities management industry, as they both begin to look beyond the construction phase of a project to the whole life of the facility. In his 1994 report on the construction industry, Sir Michael Latham recommended benchmarking to achieve the target 30% cost reduction by the year 2000 (Latham, 1994). Despite being similar in many ways to a long line of reports reviewing the construction industry, beginning with the Simon Report (1944), it was the first to begin a real change process in the industry. The Construction Industry Board was established to implement many of Latham's suggested strategies through working groups and seminars. However, Latham did not challenge the central tenets of the industry. His report did not propose a fundamental rethinking of the way in which construction takes place.

A number of other industries had met significant challenges from global competition by radically changing the way in which they worked. Seeking an understanding of what similar change might bring to construction prompted the establishment of the Construction Task Force in 1997 by the Deputy Prime Minister. Sir John Egan was asked to chair the Task Force because of his previous role in the automobile industry (as Chief Executive of Jaguar) – an industry which had already faced a challenge of this nature. The Task Force's remit was to advised, from the client's perspective, on the opportunities to improve the efficiency and quality of delivery of UK construction, to reinforce the impetus for change and to make the industry more responsive to customer's needs.

Table 18   Proposed Targets for the UK Construction Industry.

Indicator	Improvement per year
Capital Cost All Costs excluding land	Reduce by **10%**
Construction Time Time from client approval to practical completion	Reduce by **10%**
Predictability Number of projects completed on time and within budget	Increase by **20%**
Defects Reduction in the number of defects on handover	Reduce by **20%**
Accidents Reduction in the number of reportable accidents	Reduce by **20%**
Productivity Increase in value added per person	Increase by **10%**
Turnover and profit Turnover and profits of construction firms	Increase by **10%**

In 1998, the Task Force published its report, 'Rethinking Construction', which subsequently became known to many as the 'Egan Report' (1998). The Task Force identified a strategy for change that included demonstration projects, support networks and widespread dissemination of best practice. One of the key outputs from the report was the establishment of the Movement for Innovation.

The Task Force recognised that a key part of change management is the ability to measure progress towards objectives. Ultimately the Task Force suggested that an industry wide measurement scheme might be appropriate but in the short term it made specific recommendations for sustained improvement. These are summarised in Table 18. The targets were derived from a combination of existing achievements at the leading edge of the construction industry coupled with knowledge of what has been achieved in other industries. These targets are currently being developed to include the whole life performance of a facility.

The Movement for Innovation

The Movement for Innovation (M4I) was launched in November 1998 as a forum to examine progress towards meeting the targets expressed in Table 18 (URL, b). It is tasked with developing criteria for assessing a range of demonstration projects, including developing benchmarking systems, and assisting with the dissemination of best practice, as well as looking at the wider issues relevant to the Egan Report. It is assisted by the Construction Best Practice Program (CBPP) of the DETR, which is currently putting in place a best practice 'signposting service' to direct companies seeking advice on best practice to existing benchmarking initiatives such as Inside UK Enterprise (CBPP, 1998), Fit for the Future (URL, c), PROBE (1998) and the UK Benchmarking Index (DTi, 1999).

Over 80 demonstration projects have been proposed, of which approximately one third are public sector projects, with the remainder from the private sector. These projects include various framework agreement projects, as well as a variety of one-off highway, rail, bridge, nuclear and water projects. The total value is around £1.0 billion. The M4I will shortly be deciding which of these projects meet one of the four key Egan processes: product development, project implementation, partnering the supply chain, and production of components.

Within M4I, a small group of representatives from industry are also attempting to develop criteria for assessing demonstration projects, due to be published in Summer 1999.

Associated Client Initiatives

There have been a number of other initiatives by clients seeking to implement Egan's vision. Some of the most high profile initiatives have been championed by government, which is keen to be seen as a best practice client.

Within 1999, the Ministry of Defence (MoD) is likely to introduce a new construction procurement strategy known as Prime Contracting. The MoD and its chosen Prime Contractors will form a ring of strategic partnerships comprising specialist contractors and suppliers each with a role in the life cycle of a project. The system is designed to improve and develop the project brief, by evolving detailed solutions at its inception, thereby addressing project planning and constructability issues early in the project life cycle.

HM Treasury has recently commissioned a study from the Agile

Construction Initiative team at the University of Bath to assess the impact of previous procurement reviews such as Latham, and determine what is needed to provide 'best practice' public sector procurement (AC1, 1998). The Study demonstrated a 'performance gap' across the whole public sector, highlighting that a risk averse culture largely remains within public sector procurement. Although the Government has now committed itself to procurement based on value (rather than lowest price), in practice it was conceded that a non-value oriented approach still prevails. There is also still significant misinterpretation of the need for public accountability, with the perception that EC guidelines are more restrictive than is actually the case. A parallel pilot benchmarking study of 60 central government projects confirmed that two-thirds of projects ran over budget and experienced delays (PBS, 1998).

A three-year action plan is being developed by the Treasury's Government Construction Clients' Panel. Underpinning this will be a strong drive to raise awareness, and educate construction clients in good procurement practice. Those departments and bodies already applying better practice will be held up as examples for others to learn from. Early target areas for improvement include better understanding of roles and existing tools, more use of benchmarking and improvements in payment procedures. Longer-term cultural changes would be directed towards increased partnering, better management of risk and succession planning.

The Construction Round Table (a forum of twelve major UK construction procurers) and the Construction Clients Forum (which represents the majority of UK construction procurers) have recently produced documents that are closely aligned with 'Rethinking Construction'. Within 'The Agenda for Change' (CRT, 1997) and 'Constructing Improvement' (CCF, 1998) both groups have committed themselves to improving the planning, delivery and operation of their facilities through developing partnerships with their contractors. Member organisations such as BAA and Railtrack have invested heavily in partnership arrangements, and are in a stronger position to reap the rewards from partnering because they have large construction programmes. However, it remains to be seen if such term contractor arrangements can be maintained if capital investment budgets are reduced. It should also be noted that these high profile clients represent only about 15% of the industry's workload, with the remaining 85% of the industry working primarily on one-off projects that also need to be targeted if the Egan savings are to be achieved.

UK Benchmarking Initiatives

There are many benchmarking initiatives and services available in the UK, and the number is growing rapidly. Some are supported by Government, such as the PROBE Service offered by CBI, others are promoted by private organisations such as the European Construction Institute, ConstructIT and British Quality Foundation, and a number are led by academic institutions such as the AGILE Construction Initiative at the University of Bath. A few of these are described in more detail below. In addition, there are many related 'best practice' initiatives which aim to share knowledge such as the CBI's 'Fit for the Future' initiative, the Construction Best Practice Programme at the DETR, and the construction area of Inside UK Enterprise (IUKE).

Industry Level Benchmarking

ConstructIT aims to be a co-ordinating force in the application of IT within the construction process as a contribution to innovation and development of best practice (URL, a). Members comprise private companies, research organisations and academia, which use IT in the construction and facilities management sectors. ConstructIT has produced a number of benchmarking reports for different phases of the project life cycle, in which performance and level of IT usage is compared with other industries such as the automotive and shipping industries. One of these benchmarking reports is specifically aimed at the facilities management sector.

Company Level Benchmarking

The PROBE (Promoting Business Excellence) Initiative has been in existence for a number of years in the manufacturing sector. The service provides a tool for assessing an individual company's performance against a database of similar companies. This is a high level benchmarking exercise, which has recently been extended to the service sector. The model is influenced by the Business Excellence Model and Malcolm Baldridge National Quality Award Model, and has been developed in conjunction with London Business School and academic partners in the US. It examines a number of key areas within the business. These are leadership, service process, people and performance management. CBI acts as a facilitator in the benchmarking process, and provides benchmark reports to the participant.

Project Level Benchmarking

The ECI Initiative is modelled upon the US Construction Industry Institute methodology, which has been used successfully in the US for a number of years (URL, e). To date, the ECI membership has primarily consisted of petrochemicals companies, and as such its database of projects against which a new project is benchmarked is somewhat limited to the offshore industry. Project performance metrics are aimed principally at cost, schedule and health and safety performance across the different phases of a project life cycle. The ECI also measures project performance against the level of usage of a number of 'best practices' which include teambuilding, constructability, and pre-project planning.

International Initiatives – USA

The views expressed in the Egan report are not unique to the UK's construction industry. Many other industrialised nations are also examining ways to significantly improve their construction and facilities management industries. One such country is the United States, which has developed a set of National Construction Goals (NCGs) in an attempt to measure its performance. The NCGs are baseline measures of industry performance which are being developed by the National Institute of Standards and Technology, a part of the US Department of Commerce (URL, f).

The NCGs are remarkably similar to the goals proposed by Egan, and are as follows:

(1) 50% Reduction in Delivery Time.
(2) 50% Reduction in Operations Maintenance and Energy Costs.
(3) 30% Increase in Productivity and Comfort.
(4) 50% Fewer Occupant-Related Illnesses and Injuries.
(5) 50% Less Waste and Pollution.
(6) 50% More Durability and Flexibility.
(7) 50% Reduction in Construction Worker Injuries and Illnesses.

The baseline year for the NCGs is 1994, with the objective of having technologies and practices capable of meeting the goals by 2003. Baseline measures have now been developed for the first two goals described above, and work is currently underway on Goals 5 and 7 (Chapman & Rennism, 1998 a and B). This is an important first step in attempting to quantify overall industry performance, without which it is impossible to gauge the magnitude of improvements. There are a variety of public and private sector organisations in the US which are striving to meet these targets. Many of these organisations are reaching the same conclusions as Egan, in observing that the best performing projects make increased use of principles such as partnering, designing for construction (sometimes referred to as constructability), standardisation, and performance measurement (benchmarking). Within the US Government, there are many initiatives aimed at meeting improved goals demanded by the 1993 Government Performance and Results Act. Many of these are co-ordinated by the National Partnership for Reinventing Government formerly the National Performance Review, (URL, f). Private sector organisations which are promoting industry improvement include the Construction Industry Institute, and the American Productivity and Quality Centre, (URL, g).

Within the US Government the Naval Facilities Engineering Command (NAVFAC) has commissioned a variety of independent benchmarking studies, carried out by Independent Project Analysis (IPA), a private benchmarking consultant. IPA examined 32 projects for FY 93-94 and compared various results with their internal database of projects. Their output metrics (termed project success indicators) included such factors as percentage design complete at project authorisation, design change control, degree of difficulty and so forth. The results generally indicated that the Navy's Engineering Command fell short of best practice results.

The US Army Corps of Engineers (USACE) has responsibility for some 800 large civil engineering projects at any one time, with approximately 17,000 requests for work funding each year. By the year 2000, USACE expects to have to cope with many structures which have exceeded their design life (e.g., over 50% of all lock structures). In order to prioritise funding, it is developing systems for evaluating the condition of its existing infrastructure stock. One of the tools it is using is the REMR System, (Repair, Evaluation, Maintenance and Rehabilitation). REMR uses a sliding scale of condition indices, which are consistent throughout the Corps, to assess the times at which critical maintenance activities should take place. It does this by breaking down the structures into components with agreed checklists for maintenance inspection criteria. Component maintenance is weighted more strongly as the condition index deteriorates, such that high cost maintenance towards the end of the design life is minimised. Problems arise where there is no correlation between condition of a structure and its ability to perform a specific function (e.g., coastal defences and levees). Although REMR is relatively crude and to an extent subjective in the method of site assessment, it represents a move in the right direction towards uniformity within the O&M of USACE infrastructure. Similar software for the evaluation of roofs, railroads and pavements are available through University of Illinois at Urbana.

Elsewhere in the US the National Institute of Building Sciences (NIBS) is benchmarking facility performance. Their Facilities Maintenance and Operations Committee is currently producing a 'case study report' which examines 5 flagship federal facilities which have been selected as best in class in terms of operations and maintenance costs.

Other organisations in the US involved in benchmarking whole life costs of facilities include the International Facility Management Association (IFMA), the Building Owners and Managers Association (BOMA) and the Energy Information Administration (a part of the US Department of Energy). IFMA, founded in 1980, now has 15,000 members (mainly in North America), and 130 'corporate members' split into 124 chapters (8 overseas, of which 5 are in Europe). It is linked with the British Institute of Facility Management, and operates from its European headquarters in Brussels.

WS Atkins is participating in a number of initiatives that are likely to have a significant impact upon the future direction of the construction and facilities management sectors. In the UK, it is involved with a number of benchmarking and best practice programmes, including the Movement for Innovation, a number of CIRIA research projects, ConstructIT, AGILE, acting as a host within the IUKE best practice scheme, and a variety of others. In addition, it has also played a key role in the development of the US National Construction Goals. This involvement in external initiatives is an essential complement to its own internal business improvement programmes on topics such as efficient management of projects, best practice design processes, whole life costing methods and the development of integrated communication and design techniques.

Experience in these areas has also led to the management of benchmark clubs on behalf of customers. Initially, the focus was very much upon cost reduction to set benchmarks. However, the club is now looking in much more detail at the different processes involved in different MTM contracts to enable more rigorous comparisons to be made. Benchmarking initially involved cost assessment of 12 standard activities within two of these commissions in 20 regions of the country. This led to the production of league tables for contractors. However, it was recognised that this was somewhat crude, so this process was refined. In partnership with one of our main customers, we have developed a system for scoring contractors based upon service delivery, administrative efficiency, site quality, and customer satisfaction criteria. This system allows us to reduce our level of desktop audit for projects, whilst improving predictability of service delivery from our contractors. This system has now been adopted by several of our other customers, and we are looking to introduce it more widely within WS Atkins.

Conclusion

Benchmarking is an important tool to help deliver continuous improvement and spread innovation and best practice across companies and industry sectors. It is employed by many world class organisations, and championed by government and major clients. Many companies see benchmarking as a practical and beneficial business tool, all learning a great deal through involvement with the extensive range of external benchmarking initiatives that have been mentioned in this Chapter. Finally, it should be stressed that benchmarking, properly applied, is a business improvement tool, not merely a measurement and comparison technique. The biggest danger to avoid is that an organisation can end up valuing what it can measure, rather than measuring what it should value.

References

ACI (1998), *Constructing the Best Government Client*, The Government Client Improvement Study.

'Best in Class' Initiative (1998) *New Civil Engineer*, 1-8 January.

Camp R. (1989) *The Search for Industry Best Practices that lead to Superior Performance*, Quality Press.

Construction Best Practice Programme (CBPP) (1998) *Inside UK Enterprise*, DETR.

Construction Client's Forum (CCF) (1998) *Constructing Improvement - The Client's Pact with Industry.*

Chapman R. & Rennison R., (1998 a) *An approach for Measuring Reductions in Operations, Maintenance and Energy Costs : Baseline Measures of Construction Industry Practices for the National Construction Goals*, NISTIR 6185, National Institute of Standards, USA.

Chapman R. & Rennison R. (1998 b) *An approach for Measuring Reductions in Delivery Time: Baseline Measures of Construction Industry Practices for the National Construction Goals*, NISTIR 6189, National Institute of Standards and Technology, USA.

(CRT) Construction Round Table (1997) *Agenda for Change.*

(DTi) Department of Trade and Industry (1999), *The UK Benchmarking Index.*

Egan Report (1998) *Rethinking Construction*, Report of the Construction Task Force, DETR, HMSO, London.

Latham Report (1994) *Constructing the Team-Final Report*, HMSO, London.

PROBE (1998), *Benchmarking: The Probe Family-promoting business excellence,* CBI, London.

Simon E. D. (1944) *The Placing and Management of Building Contracts.*

(URL, a) refer to http: www.npr.gov

(URL, b) refer to http: www.m4i.org.uk

(URL, c) refer to http: fitforthefuture.org.uk

(URL, d) refer to http: www.constructit.salford.ac.uk

(URL, e) refer to http: www.info.lboro.ac.uk
(URL, f) refer to http: www.nist.gov
(URL, g) refer to http: www.npr.gov
(URL, h) refer to http. www: construction-institute.org

25 Change Management through Design

Christine Landorf and Annemarie Harrison,
University of South Australia

The discipline of facility management has developed in response to an increasing business sector awareness of the value of capital investment in facility assets and the impact that maintenance costs, user satisfaction and changing technology can have on organisational performance. As an extension of this, many organisations have been aware for some time that workplace design is often indicated as a key factor in the level of user satisfaction and productivity (Smith & Kearny, 1996).

This chapter describes the results of a pilot research project undertaken by the University of South Australia. The project involved the analysis of organisational information and data, and the measurement of the physical settings of two organisations undergoing office relocations. One organisation was a State Government Department in South Australia referred to as Study 1, and the other a major State Government service provider presently undergoing part privatisation, referred to as Study 2. The two studies measured the physical settings and user satisfaction levels of several departments within each organisation. A variety of additional historical, structural and human resource information and data was also collected and analysed. The research provided a preliminary framework for the early identification of key design parameters in effective office accommodation with the view to enabling more predictable relocation outcomes for organisations.

Context of the Research

Measuring performance in the workplace is a very complex issue in its own right (Akin & Hopelain, 1986; Lloyd, 1990; Larson & Callahan, 1990; Brill, 1992; Brown & Mitchell, 1993; Leaman, 1995) but proving that there is a direct link between performance and workplace design is further complicated by the inter-dependency of performance and issues such as organisational design and culture (Handy, 1985; Barney, 1986; Nadler & Tushman, 1988), leadership style and change management (Simon, 1956; Dunphy & Stace, 1988 & 1993, Dunford, 1990), building design and design briefing (Duffy, Laing & Crisp, 1992; Bordass & Leaman, 1996), and internal environmental control (Leaman, 1992; Leaman & Bordass, 1993 & 1997). The need to challenge current theory and practice in regards to the relationship between performance and workplace design is considered vital to balance the present focus of facilities managers on asset and building management as a means to improving the bottom line.

Evidence of the link between work environment and user behaviour can be examined on a global scale through publications, conference presentations and professional dialogue between international academic researchers and consultancy organisations in both the private and public sectors. The United Kingdom and the United States of America figure prominently in this research arena while Australia is also developing a healthy FM research culture of its own.

Intervening Factors

Organisational management influences organisational behaviour and user satisfaction in work settings (Bordass & Leaman, 1996). It is important to recognise that individual differences in management styles can create a dilemma in the interpretation of research results as can ongoing or abrupt organisation change. Case studies are used in research because they reflect the complexities of real life. These complexities therefore need careful identification.

Work Activity Settings

Typical work settings within organisations support a number of work activities or types. These different work and communication activities can impact on an organisation, as can the designation of functional spaces to support them. These activities include individual work, collaborative work, work related communication and social interaction. Communication may also occur between two or more persons in an organisation, with or without face to face contact. E-mail, voice mail,message bank and teleconferencing are all examples of increasingly favoured communication modes.

Collectively, the investigation of the whole of a corporate organisation should give results that are greater than the sum of all of the parts. Such parts include function of the designated personal and shared spatial resources (intended use, use in practice, etc.), and an accessibility factor. The notion of accessibility can be expanded to include control over access to space and communication, as well as individual control over personal space and the surrounding conditions of the work setting (Leaman & Bordass, 1995).

User Issues within an Organisation

Members of an organisation include female and male staff, as well as members with various lengths of service and levels of commitment to the objectives of the organisation. Such staff characteristics and differences are often important factors when interpreting the results of a study. There are also indications that perceptions of comfort may be linked to job design and perceptions of individual autonomy within an organisation. Rowe (1993) suggests an association between less active tasks and dissatisfaction with climatic conditions whilst Leaman and Bordass (1993) suggest a perceived lack of control over internal environmental conditions may be associated with dissatisfaction.

Whilst organisational changes, such as outsourcing and downsizing, can impact on staff attitudes, the concern of certainty in employment as a result of new business plans or weak leadership can lead to high rates of staff turnover and illness (Leaman & Bordass, 1995). Sometimes a new location/workplace is simply novel, and that novelty or unfamiliarity can, when combined with other factors such as length of service, impact on a staff member's response to a work setting (the Hawthorne effect). Some respond well to a new situation and what is perceived to be increased management attention whilst the result in others can be a higher level of stress, particularly if they are not used to change.

Research Method

The two studies aimed to:

- Use conceptual frameworks and research methods drawn from psychology, building and management science to assess the impact of the design of the physical work environment on organisational communication;
- Test criteria for the measurement of the links between the physical environment, human and organisational behaviour and productivity;
- Determine key criteria which will enable more predictable relocation outcomes for organisations involved in the design of office accommodation and relocation of organisations and subgroups, and
- Consider the relationship between the design and implementation process, and the management process by assessing organisational communication and efficiency.

Following the identification of a number of departmental work settings within the participating organisations, each with a similar management structure and members who had been relocated from traditional cellular office environments to new open plan settings, the research was structured as follows:

- Physical Surveys involving quantitative measurement of internal office environments including space standards and planning, functional relationships and physical characteristics through both document searches and physical measurements.
- Descriptive Surveys of internal environments including building fabric characteristics and wear patterns through photographic and video records.
- Self Report Surveys of users responses, comparing aspects of the new and old accommodation including ambient conditions, building services and the psychological environment through questionnaires based on User Benefit Design Criteria (Murtha, 1976).

The research outcomes were expected to lead to the development of models that related organisational characteristics, specifically productivity, to the physical environment.

Research Results

The self-report survey consisted of a seven point Likert scale used to measure levels of individual satisfaction and dissatisfaction with a number of key environmental criteria. Several open ended questions were also included in the questionnaire. Responses to both studies were initially categorised into those that expressed overall satisfaction with their environment across all criteria, those that expressed overall dissatisfaction and those results that did not indicate either satisfaction or dissatisfaction. The results are shown for both studies in Figure 49 and Figure 50.

Figure 49   Study 1 User Satisfaction Levels.

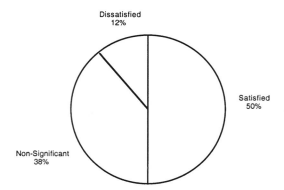

Figure 50   Study 2 User Satisfaction Levels.

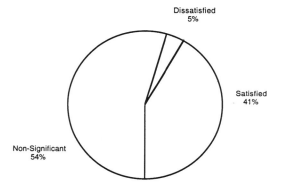

Study 1 surveyed four diverse sub-groups or departments within the same organisation. Of these, three were dispersed over two floors (Floors A and B) and one was located on a third floor (Floor C) within the same building. This contrasted with the wider disbursement of organisational sub-groups in Study 2. Again four sub-groups were surveyed, but two of these occupied separate levels in the same building (Floors W and X) and the other two were accommodated in a further two buildings (Floors Y and Z). It is interesting to note that one of the latter groups occupied a building some distance away from senior management. Satisfaction levels for the various questionnaire criteria varied within each sub-group. The significant variances in either satisfaction or dissatisfaction for the various criteria are shown in Tables 19 - 22.

One sub-group in Study 1 were predominantly female and with 1 - 5 years service in the organisation. The most negative responses related to Climatic Conditions which cited erratic air temperature, poor air circulation and poor air quality. Yet this was the most positive overall in most other categories. The most negative responses were expressed by the two sub-groups on Floor A, predominantly male, with more than 5 years of service with the organisation. Table 20 indicates the responses of this sub-group which contradict other population responses.

Table 19　　Study 1 – User Responses to Questionnaire Criteria.

Positive	Negative
Work Facilitation Acoustic Environment Hazard Free Work Setting Awarness of External Environment Ownership of Work space Pleasant Work Setting Opportunity for Interpersonal Interaction Work Space Improvement	Climatic Comfort Workplace Environmental Conditions

Table 20　　Study 1 Floor A – Criteria Responses that Contradicted the Study Population.

Questionnaire Criteria
Work Facilitation Environmental Stimuli New Work Setting Enables Task Focus Pleasant Work Setting Work Space Improvement Space Planning for Interaction/Access Paths Design for the Disabled

Some of the open responses highlighted dissatisfaction with Climatic Comfort; Ambient Conditions; Hazards as well as Design for the Disabled. These criteria were further cited in relation to specific problems such as cleaning/maintenance and inappropriate design for some of the public functions.

Figures 49 and 50 show that there were similar patterns of overall satisfaction levels between the two studies. At the individual criteria level however greater concern about noise interference was indicated in Study 2. Females mirrored the responses of Study 1 and negatively responded to Climatic Conditions and Workplace Environmental Conditions. More males in some sub-groups expressed dissatisfaction and impassioned open responses were a significant feature of Study 2.

Table 21　　Study 2 – User Responses to Questionnaire Criteria.

Positive	Negative
Work Facilitation Hazard Free Work Setting Awarness of External Environment Ownership of Work space Pleasant Work Setting Opportunity for Interpersonal Interaction Work Space Improvement	Climatic Comfort Workplace Environmental Conditions

Differences in the response profiles in Study 2 indicated a possible disparity in management style across the four sub-groups. Management style in Study 1 appeared to be relatively homogeneous however with differences in responses relating to control of work and climate as well as variables such as gender and tenure.

Table 22   Study 1 Floors W and X – Criteria Responses that Contradicted Study Population.

Questionnaire Criteria
Improved Task Focus Provision for Professional Interaction Provision for Privacy Work space Improvement Space Planning for Interaction/Access Paths Design for the Disabled

Overall, Study 2 responses were the least negative and the least positive across most criteria. The two sub-groups housed in a building with an air-conditioning intake at street level in the congested central business district expressed the most negative and extreme responses on air quality. One of these groups with stair access only to top level office accommodation also responded negatively to disability access. The belief that the accommodation did not adequately facilitate work is also apparent in the comments expressed about the lack of space provision for work activities and the inability to sense the time of day. A profile of this sub-group indicated a large percentage of long tenured professionals engaged predominantly in field work activities.

Findings

Respondents in Study 1 generally found space planning in the new accommodation to be more legible and integrated, and the environment brighter and more comfortable than respondents in Study 2. The space planning attended to the needs of staff to some degree enabling flexible work practices such as group/team discussion areas. Negative responses related to the lack of user consultation on provision of some key facilities. In one sub-group of the organisation personal safety was also of some concern. There was also a more positive response from users, senior management and staff to the newly designed accommodation where the designers were informed on matters relating to specific user needs. When this dialogue failed to address key functions, the resulting design was seen as inadequate for the needs of staff/users in those workplaces and staff were more negative and less satisfied with their work space.

Attitudes to design consultation were generally mirrored in Study 2 with one exception. One sub-group, Floor Y, engaged professional strategic planners and designers throughout the relocation process, and actively communicated with staff members in the consultation and design process. In this case the degree of satisfaction was higher with no dissatisfied responses.

In both studies, open-ended questions initiated emotional responses in regard to cleaning and the quality of the air-conditioning. Issues of intrusive noise, lack of acoustic privacy and privacy in relation to spatial planning were also cited. Early analysis of the corporate data associated with Study 2 also indicates some relationship between work satisfaction and management style. Where management of change is integrated through effective communication organisational members appear to be more satisfied in the workplace.

Discussion

There is a dilemma in all fields research of being able to isolate the phenomena being studied. To separate management from organisational behaviour and the work environment is almost impossible for any organisation. Organisational data such as a high level of sick leave taken over a critical period of organisational change could indicate staff dissatisfaction with that change, but could equally relate to an influenza epidemic within the same time frame. It is through replication of a large number of studies such as this that a more definite picture will emerge but certainly several issues have been highlighted by the two studies that suggest areas for further serious research.

Of these, management style and culture at the sub-group or departmental/divisional level is one issue that has been suggested as significant in terms of its potential impact on the level of satisfaction. The differences in sub-group culture were more evident in Study 2 where the various departments involved have traditionally operated with a reasonable degree of autonomy along distinct professional lines. Variations in the results between the two studies start to suggest that this is one area that requires further research specifically to determine a measure for organisational culture type and capacity for change which will in turn inform the briefing process.

An extension of this is the impact that leadership style and organisational structure at the sub-group level have on communication, motivation and change management during a relocation process. Leadership has the propensity to either block or facilitate change within any organisation, typically through communication processes. Leadership style has been well studied as has organisational design (Ott, 1996) but the potential impact of these two critical areas of behavioural study upon the accommodation briefing process has not been adequately tested.

Almost all organisations undergo constant change but there was difficulty in finding a relatively stable organisation to commence this research. A major state government restructuring was occurring in South Australia at the time of Study 1 and several earlier attempts to commence the research had to be aborted due to the impact of organisational change. This had some associated negative effects in relation to user attitudes and overall work related performance indicators. Both the co-operating organisations and the researchers where aware of intrusion, albeit minimal, when members of an organisation are in fear for their jobs.

Conclusion

A single coherent and focused approach, such as strategic facility planning, could improve work place satisfaction. Seidman (1997) suggests facility planning has an opportunity to singly impact the morale of organisational members via the facilitation of pleasant and properly supported facilities despite external and internal influences. Several key issues have emerged from these studies in relation to satisfaction in the workplace however. At this early stage these can only be broadly identified. Further testing is required to confirm the reliability of both the research method, and the consistency of these key issues across a number of organisations and work settings. The principle conclusions are that satisfaction in the workplace, and hence performance, may be influenced by:

- Length of service in an organisation impacting on the capacity of individuals and groups to accommodate change;
- Job design and level of autonomy in work activities;
- Degree of personal control over environmental conditions;
- The management of change in an organisation including communication and consultation with staff and leadership style;
- The level of staff consultation and involvement throughout the relocation process;
- The impact that design has on facilitating communication within an organisation.

In future studies of this type, the measurement of performance needs to be considered in greater depth as does the influence on performance of organisational structure, management style and change in both the external and internal environments of an organisation. There is a general need to view all of the issues impacting on organisations and productivity from a much broader framework. Such a framework should consider an organisation to be an inter-related whole greater than the sum of its parts. For designers, this means extending the focus at the briefing stage to include less tangible issues such as corporate culture and individual world views, in addition to function and cost.

These conclusions echo what has been previously stated by Bordass and Leaman (1996) who borrow from Checkland and Scholes (1990) when they suggest that ". . . stable, self-sustaining, workable and efficient systems have four basic properties: hierarchy, integration, communication and control." Systems methodology (Checkland and Scholes, 1990) provides a way of modelling a variety of complex and ill defined issues, not immediately associated with a design project. This allows the four inter-related properties of any system to adapt and change with their environment without destroying the whole.

These studies point broadly to control as being a significant contributor to satisfaction in the workplace and hence system efficiency. Communication is also indicated as a key factor. Future research will concentrate on these issues in relation to the design brief and the briefing process, and in relation to control at both the macro level of strategic planning, and at the micro level of the individual.

In conclusion, studies as outlined in this chapter are of significant long term value to organisations as they provide:

- An independent and non-threatening forum for organisation members to express their attitudes (Shipley, 1991);
- Research methods which are structured to achieve consistency and reliability in results; and
- A clear and structured picture uninfluenced by the values and perceptions often found within organisations (Cohen et al., 1997).

Fashionable words such as 'hot desking' have found their way into the popular press (Newman, 1997) and have become symptoms of organisational change rather than remedies for deeper economic and social issues in relation to the workplace. Woudhuysen (1997) argues against the fantasy, beautiful dreams of 'virtual offices' which fail to address larger organisational issues such as structure, culture and management style. These types of social experiments are often based on shallow research and tend to contradict conventional tested wisdom in the fields of psychology and management science. Office accommodation is a significant investment for any organisation and it deserves to be undertaken in the knowledge that better performance will in fact result from better design.

References

Akin G. & Hopelain D. (1986) Finding the Culture of Productivity. *Organizational Dynamics*.

Barney J. B. (1986) Organisational Culture: Can it be a source of sustained competitive advantage? *Academy of Management Review*.

Bordass B. & Leaman A. (1996) Future Buildings and Their Services: strategic considerations for designers and their clients. *CIBSE Conference*.

Brill M. (1992) Workspace Design and Productivity. *Healthcare Forum*.

Brown K. A. and Mitchell T. R. (1993) Organizational Obstacles: links with financial performance, customer satisfaction, and job satisfaction in a service environment. *Human Relations*.

Checkland P. and Scholes J. (1990) *Soft Systems Methodology in Action*, John Wiley & Sons Ltd, London.

Cohen R., Ruyssevelt, P. & Standeven M. (1997) The Probe Method of Investigation, *Probe Methodology*.

Dunford R. (1990) Discussion Note: strategies for planned change. An exchange of views between Dunford, Dunphy and Stance. *Organisational Studies*.

Dunphy D. & Stace D.(1993) The Strategic Management of Corporate Change, *Human Relations*.

Dunphy D. & Stace D.(1988) Transformational and Coercive Strategies for Planned Organisational Change: beyond the O.D. model, *Organizational Studies*.

Duffy F., Laing A. & Crisp V. (1992) The Responsible Workplace, *Facilities*.

Handy C.B. (1985) *Understanding Organisations*, 3rd. ed., Penguin, Harmondsworth.

Larson J. & Callahan C. (1990) Performance Monitoring: how it affects work productivity. *Journal of Applied Psychology.*

Leaman A. (1995) Dissatisfaction and office productivity. *Facilities.*

Leaman A. (1992) Open-plan Offices: kill or cure. *Facilities.*

Leaman A. & Bordass B. (1997) Strategies for Better Occupant Satisfaction. *Fifth Indoor Air Quality Conference.*

Leaman A. & Bordass B. (1993) Building Design, Complexity and Manageability. *Facilities.*

Lloyd B. (1990) Office Productivity: time for a revolution? *Long Range Planning.*

Murtha D. M. (1976) *Dimensions of User Benefit*, Washington D.C.: American Institute of Architects.

Newman G. (1997) Dare to Hot Desk, *The Weekend Australian*, August 30-31.

Otts J. (1996) *Classic Readings in Organizational Behaviour*, 2nd ed., Harcourt Brace College Publishers, Orlando.

Rowe D. (1993) Health Risks: Health Hazards in Buildings, *Hazards.*

Shipley P. (1991) The Analysis of Organisations as a Conceptual Tool for Ergonomics Practitioners. In: *Evaluation of Human Work: A practical ergonomics methodology*, (Ed. by J. R. Wilson & E. N.Corlett.) Taylor and Francis, London.

Seidman R. (1997) Implementing Facilities Planning and Asset Management in Universities, *FMA Australia 97 Conference.*

Simon H. (1956) Rational choice and the structure of the environment. *Psychological Review.*

Shwartz H. & Davis S.M. (1981) Matching corporate culture and business strategy. *Organizational Dynamics.*

Smith P. & Kearney L. (1996) *Creating Workspaces Where People Can Think*, Jossey-Bass, San Francisco.

Woudhuysen J.(1997) The Future of Space, FMA Australia '97 Conference

26 Intellectual Capital: future competitive advantage for FM.

Peter McLennan, University College London

How can facility managers translate their unique operational expertise, within the globally competitive business environment, into value added knowledge? This question is now challenging both facility managers and the organisations for which they work. There are two reasons why knowledge has become a strategic management issue. First, knowledge can now be captured and structured explicitly through computer based management systems. Second, in order to remain competitive, businesses must make informed decisions on product and service offerings more quickly (DTI, 1999). So the active management of knowledge within a firm has become an explicit rather than implicit requirement, and is seen therefore, as a the strategic issue for the future. The challenge for the facility manager now, is how to transfer their expertise into 'packaged useful knowledge' and so deliver a tangible business asset. This 'packaged useful knowledge' may be considered as equivalent to the concept of 'intellectual capital' in other fields (Galbraith, 1973) and (Stewart, 1997). Whether facility managers will become part of this knowledge revolution will depend on the degree to which they perceive these developments as an opportunity or as a threat.

For the facility manager the opportunity is to demonstrate that their operational expertise and data is of direct relevancy to an organisation within a particular business sector. For example, aspects such as occupancy costs, energy consumption, density of use, asset appreciation and utilisation rates of the physical resource together with their methods and procedures for the effective management of facility resources, will all need to be transformed into 'useful knowledge packs'. The threat of inaction here, is the loss of FM's competitive position by allowing others to acquire, manage and exploit this knowledge and with it the control of the FM resource area. The commercial office sector, which constitutes a physical resource with both product and services operated by facility managers, has a complex set of customers. While a customer focus remains problematic for many building professionals (Latham, 1994), the term reflects an underlying quality management framework used currently by most businesses. In fact, much like manufacturing, the actual external or paying customer (consumer) enters the premises rarely, so for facility managers the customers are predominantly internal. As a result, within the commercial office sector, the direct business benefits of facility management knowledge may be less clear than in other sectors such as retail and leisure.

The procurement of the commercial office product involves complex relationships between owners, investors, leases and clients. The information flow for a typical project is illustrated in figure 51. The customer making the decision about financing an office project if a developer (finance box), may be selling this on to an investor prior to tenant occupation (operate box). Further, the briefing documents that informed the design may be irrelevant to the future generations of occupants, who constitute the 'real' internal customers.

Given this reality, the future of FM knowledge development in the context of the office sector, appears to be taking two divergent paths. One path for the FM provider is to focus on the physical resource as an investment product, with the 'customer' as the owner of this investment. This path follows the need to provide an FM service that maintains the asset over time for the investor with the knowledge system developed around the traditional view of property as a financial asset. The second path focuses on the business and their internal 'customers' (i.e. their employees) who use the office. Here the knowledge system is service based with a potential to change the facility manager's role, similar to that of the operations manager in manufacturing, with the knowledge system aiming to support the processes and operations of the business directly.

Figure 51 Information flows for physical resources.

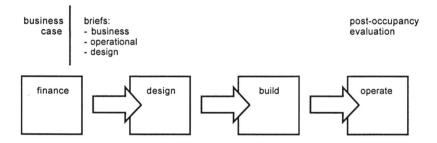

This chapter develops the concept of 'intellectual capital' in FM along these two alternative paths. The potential benefits of knowledge management are examined in relation to both the commercial office and retail sectors. It concludes with a speculative account of the opportunities that knowledge management could create for facility management, as well as the benefits that it promises to provide for business.

The Knowledge Context

The classification of intellectual capital begins with the proposition that "knowledge assets, like money or equipment, exist and are worth cultivating only in the context of strategy" (Stewart, 1997), with business strategy choosing where to place resources within a market segment (Vasconcellos, 1999). For the facility manager of an office, the specific task is to decide how to allocate FM resources to meet the service needs of the client at a particular location. These resource allocation decisions, and the criteria on which they are based, will differ across the spectrum of organisational types, from banking to insurance, and from advertising to e-business. Here, the facility manager's unique understanding of what is required to support each of these different market segments provides the potential competitive advantage for FM.

Three categorisations are in common used in discussions related to knowledge and its classification. These are data, information and knowledge (Davenport & Prusak, 1998 and Zack, 1999). From the facility managers perspective, data are discrete facts about the physical resources which could include annual energy consumption, annual occupancy costs, number of employees per square meter, room utilisation rates and so on. These data are kept and organised into information by facility managers in order to provide businesses with management reports on both the condition and capacity of their physical resources as well as the process by which their assets are being maintained in support of business activity. The facility manager's knowledge is both object related - for example understanding the electrical systems in buildings - and process related - for example, procedures for health and safety or for undertaking risk assessments. The specific facility management knowledge that has strategic value, is the understanding of the relationship between the performance of the physical resources and their impact on the customer being served by these resources. This type of knowledge can be difficult to access since it is often tacit and experiential. However, the intellectual capital concept provides a framework for dealing with this area of knowledge.

Intellectual capital is said to include three knowledge areas - human, structural and customer. Human capital represents the "capabilities of the individuals to provided solutions to customers" (Stewart, 1997). Structural capital comprises the operating systems, physical resources, renewal capacity and work processes of the business. This area is most closely associated with the work of facility managers. The customer capital is the relationship with the present customer base, looks at market penetration, market coverage and customer loyalty (Stewart, 1997 and Edvinsson &Malone, 1997). This framework is currently used to organise business knowledge in a purposeful way (Baladi, 1999), with the interplay between the three areas seen to creates added value and thus the intellectual capital of a business. So this model can be used to measure performance within a business. Its prime aim is to account for the 'off balance sheet' assets in annual reports and is not without its critics in the business world (Davidson & Prusak, 1998). However, this framework provides a basis and opportunity for facility management to begin to consider what their services offering might be in a few years time. The retail sector offers some pointers here, as to how knowledge can be used to add value to business.

FM Knowledge in the Retail Sector

At present, the large UK retailers like Sainsbury's, M&S, Tesco and MacDonalds are perhaps the best reported 'facility' knowledge management practitioners. They appear to manage both the 'product' knowledge related to the building specification and construction and the 'process' knowledge related to project and operations management, the former while under development and the later in use. In addition, these corporations manage the process of financing, designing, constructing, operating and monitoring their physical resources with clear customer focused business objectives. The explicit quality management approach ensures not only the potential of continuous improvement in all aspects of these projects, but also potential innovations as well. MacDonalds' innovation in pre-fabrication that allows a new restaurant franchise to open in three weeks, once planning permission is granted, is a well known example (Barlow, 1997). It illustrates the knowledge management component at both levels, object and process. The technical requirements are clearly based on the detailed operational procedures for a franchise, in itself a broader business knowledge procedures competency. These procedures have clear requirements set-out in the briefing documents. Any changes to the brief requires a demonstration of the likely performance improvement for business operations, i.e. faster service delivery or less time between tasks. From a quality perspective, the customers are very clearly in focus, the internal ones creating and delivering the product to the external ones who buy the product, all of whom are supported by the physical resources of the site, building, equipment and furniture. By structuring and managing the data, information, and knowledge flows, Mac Donalds continues to lead in their market segment. The business benefits of such an innovation are obvious. In the retail sector, facility management couple with operations management, provides important feedback between operations, finance and the business plan.

Adding-Value through Feedback

Figure 52 illustrates this generic project process, showing the flow of information and the feedback loop that is critical in understanding where the knowledge value lies for facility managers. The benefits of this managed approach appear to be:

- Clear control of project time, costs and specification;
- Awareness of the implications of decisions made through-out the process;
- A firm understanding of the customers requirements, and
- Direct performance feedback to the business plan.

Figure 52 Information flows for physical resources.

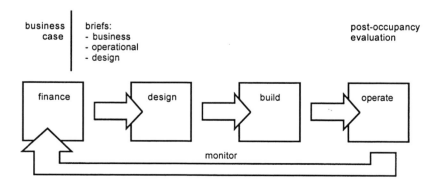

The model shown in Figure 52 has four discrete stages, shown as boxes, typically undertaken to provide a physical resource for a business, institution or government body. These stages represent the process by which a project is transformed from a business need into physical reality. The four stages are: 'finance', 'design', 'build' and 'operate'. The arrows moving from left to right indicate the information flows from one stage to the next, each of which represents a package of information that enables the next stage. The arrow moving from the 'finance' and 'operate' boxes represents the information feedback loop, essentially the formalised monitoring of business objectives. This feedback loop allows measurement of the business objectives and criteria against the physical resource performance. As mentioned earlier, this feedback loop provides details of how the physical resource is performing against the business requirements set-out in the 'finance' stage. It is the facility management professional that provides this information in order to close the feedback loop. Furthermore, the knowledge about the actual use and operations of this resource is typically known only to the facility manager.

The 'finance' stage involves both identifying and defining the problems that are to be resolved for the business through the construction of a physical resource. An important part of evaluating a project is the criteria that are set out in the project's business case. This is often a restricted document with a limited circulation. Typically, both the directly employed and contracted facility managers are unlikely to be involved in formulating this document. The project requirements are articulated through a series of briefing documents that are produced for external use. These are: business brief - articulating the goals and objectives to be obtained through the physical resource; operational brief - reviewing how the physical resource is intended to support the business goals and objectives; and design brief - outlining the key parameters of the design of the physical resource based on the business and operational briefs. This briefing information is passed to the 'design' stage. The completion of this phase of a project typically indicates a decision has been made to proceed.

It should be noted however, that the use of operational knowledge about physical resources as a means of informing design at the strategic level is not routinely reported to any great extent at present in the UK. Although the MacDonalds example has an operational brief in principal if not in fact. It can be argued that partnering works because it is able to, as in the case of MacDonalds, create a standard building product for a particular business need. However, the contemporary briefing guidance has dropped this important aspect of the cycle, of which the Construction Industry Board (1997) is a typical example. The development of the operational brief is one means of translating the facility managers information base into a knowledge base that can be used to add customer value (McLennan and Cassels, 1998) and (Salisbury, 1991) and (Nutt, 1993). The operational brief could help to make the organisational requirements clearer for the subsequent design responses in support the business activity.

The next two stages, 'design' and 'build', are the focus of attention of both the popular press and government. The recent construction industry reports by Latham and Egan and the resulting key performance indicators are two notable examples. These were in part an attempt to bring a wider customer perspective to an industry that was previously, and still is in some areas, excessively 'supply-side' focused. The 'design' stage involves a team of building professionals - architects, engineers, and surveyors - as well as product suppliers and manufacturers, who develop the design brief into a set of contract documents - building plans and specifications - which detail the specification and arrangement of materials. This design information is executed in the 'build' stage, involving the construction of the physical object - building, streetscape or motor way. The built object is then put to use to support the 'operate' stages throughout the facility's life span.

The 'operate' stage concerns the long-term occupation or use of the physical resource, under a traditional institutional leasehold for 25 years, or if freehold until the need to sell. The traditional institutional lease ties an organisation to a particular location for a very long time. In fact, the UK leasing arrangements are the longest in Europe. In the Information Age with 1-3 year business cycles, the negative effects of this could be a critical hindrance to the future development of the Knowledge Economy in the UK (Crosby and Lizeri, 1998). This stage is the knowledge domain of facility management professionals.

The model shown in Figure 52 helps to illustrate the reason for the successful exploitation of the information flows, feedback loops and resulting business knowledge in the retail sector. As discussed previously, within the retail sector there is a clear business objective for the building - to provide the environment in which a sale can take place with a customer (although the recent on-line shopping service offered by Tesco's is changing this to the distribution of customer orders). The buildings are owned or are on long leases. For large retailers, this process forms the basis of a continuous feedback loop which often results in business innovation.

In the commercial office sector the knowledge about the building's contribution to business performance remain under-developed. The formal use of post-occupancy evaluation (POE) and building evaluation techniques to provide feedback on building performance have been used with varying degrees of success for some time (Preiser, 1989) and (Baird et al., 1995). However, these tend to be discrete studies of buildings at a single point in time, focusing on the building's design and user satisfaction, not on business operations or the degree to which actual business performance matches initial expectations (Cohen, et al. 2000). Leaman and Bordass (1999) and Aronoff and Kaplan (1995) are two exceptions.

It is through the creation of a feedback loop shown in Figure 52, that the future opportunities of facility managers in the office sector lie. This is where they will have most to contribute. Ideally, the facility manager's knowledge contribution should be made at the briefing stage, but this rarely part of the current project process. At present, there are few building professionals that can realistically impart such knowledge to provide benefit to the business and their customers. The development of this knowledge of performance in use is the opportunity for facility managers to retain control of the physical resources in the future. As introduced earlier, the proposition is that based on the feedback model, there are two divergent paths for the intellectual capital of facility managers to follow. One path is for the facility management provider to focus on the physical resource as an investment product and the customer that owns this investment. The second path is to focus on the customer using the office and the supports that the core business needs.

Intellectual Capital to Investors

The first knowledge bridge for facility managers in the office sector, links operational knowledge directly to the commercial owners; that is the developer, the investor, the property group, the pension fund or the financial institution. These treat the building as a product. The benefits of this are apparent. In each building, cost performance can be systematically reported, understood and improved from a life cycle perspective. This is an important criteria for the occupiers. In addition, a well maintained and managed estate could lead to the appreciation of asset value. One contemporary example of this taking place is Stanhope Properties which created Broadgate Estates to manage the Broadgate site. The development is now owned by another property group. However, Broadgate Estates have used their intellectual capital to manage other prestige offices in London. This is based on the use of their knowledge of how to maintain the assets value for the prestige office market segment.

Another benefit gained from 'facility in use' knowledge is to understand the whole-life cost and risk position over time. The growing importance of this was introduced in Part One of this Book and has been emphasised in a number of recent publications on the Public/Private Partnership (PPP) Initiative (CIC, 2000) and (HMT, 2000). Recent experience indicates the construction costs for offices represent between 10-15% of the total project costs, the remaining 85-90% being the operating costs. Further, the information on maintaining asset value over time by understanding both the performance in use and cost in use of an office building is clearly deficient from the financial community's viewpoint

From the investor's perspective, the facility manager's ability to comment and advise on the performance objectives, as set-out in the initial finance stage, can provide tangible benefits. Two examples will be cited. The first concerns the expanded service offerings of Jones Lang LaSalle and CB/Hillier Parker, integrating facility management with commercial real estate practice. Here the connection is through service provision. The first organisation conducts the transaction - leases or sells the property - the second adds the option of providing facility management services. The second example is the serviced offices provided by Regus and HQ where the business integrates all three services - investment, transaction and operations in a single product offering.

In all of these types of development, companies are setting out to exploit the facility knowledge base. This should lead to much more significant improvements and innovation for commercial offices, than will the implementation of the recommendations from either of the construction based Latham or Egan reports. This is because the position of the facility manager has more of an opportunity to influence the design and construction of buildings by understanding how they are operated then either of the former.

Intellectual Capital for Businesses

The second bridge to make use of facility knowledge is more complex as it relies on novel responsibilities for facility managers to create the knowledge linkage through the new role of the service operations manager (Hope and Muhlemann, 1997) and (Schemenner, 1995). What makes this role possible now is the speed of business response needed in the office environment as a result of the extended use of new information and communications technologies (Castells, 1996). The transactions of many organisations are reliant on this technology. Further, the space required to support the work activities change with greater speed then before, 'e-time' being the popularised term for the shortening of the business cycle. Business activities based on teams have typically 3 - 6 month project lives and space is often re-configured to meet these requirements. In addition, the customer is able to inhabit the space 'virtually' through internet portals or call centres. This results in many office workers interacting directly with the outside customer more than ever before. The support services to these new business processes in terms of physical resources; building, equipment, furniture, site and now infrastructure, are more 'business' critical then they were previously. For the facility manager this provides an opportunity to move into a core function position as part of the operations team in a similar position to the operations managers in manufacturing facilities.

The knowledge of facility performance in use is a key element in developing strategies for supporting these new work activities. This includes an understanding of both the technical issues - network capacity, secure communications channels and call routing designs - and the process requirements - responsive procurement schemes, secure management methods for supporting a variety of infrastructure and the means of measuring occupancy performance. At the moment this development can be seen in the larger facility management providers in the UK such as Johnson Controls, WS Atkins and AMEC. However, the service operations manager's role seems to be best developed from within a business, in order to provide the continuity between the business and their physical resources. The use of the intellectual capital model with its human, structural and customer capital dimensions will allow the facility manager to contribute more directly to the service delivery than was possible before. While speculative at the moment, this is perhaps the most interesting opportunity for facility managers to contribute more strategically to the business.

FM's Knowledge and the Future

The proposition that has been outlined in this chapter is that the current facility management approach is constrained from exploiting the value of its knowledge base. There are several reasons for this, but principally it is because the feedback loop in Figure 52 does not yet exist. This is partially a result of facility management services being seen as an expense within businesses and partially a function of split property and facility management roles and responsibilities. In either case, feedback remains elusive and any knowledge gained is dispersed quickly between project participants. Another related constraint is the non-core activity label attached to facility management. This is perhaps the more telling in so far as it also suggests that the information feedback loop does not have a natural home within a non-property company. It is this emerging business development that provides an opportunity for facility management to exploit fully the knowledge gained through operating buildings, particularly office buildings in the first path. For the second, it is the continued collapse of the time frame of ordinary business cycles. This creates an opportunity for facility managers to become more involved with the actual business processes as service operations managers. This role remains to be developed. The use of information technology is the means by which the knowledge of facility managers can be captured, structured and used to enhance business value. In order to retain control of their natural areas of expertise, facility managers need to move swiftly to secure this exciting opportunity for the future.

References

Aranoff S. & Kaplan A (1995) *Total workplace performance.* WDL Publications, Canada,

Baird G., Gray J.; Isaacs N.; Kernohan D. & McIndoe G. (1995) *Building evaluation techniques.* McGraw-Hill Companies, Inc., UK.

Barlow J., Cohen M., Jashapara A. & Simpson Y. (1997) *Towards Positive Partnering.* The Policy Press, Bristol.

Barrie G. (1999) Treasury chief plans revolution for PFI. *Building,* vol. 8 January pp.9.

Castells M. (1996) *The Rise of the Network Society.* Blackwell Science, London.

Cohen R., Standeven M., Bordass W. & Leaman A. (2000) Assessing building performance in use: the PROBE process. *Building Research and Information,* forthcoming.

Construction Industry Board (1997) *Briefing the Team: a Guide to better Briefing for Clients.* Thomas Telford, London.

Construction Industry Council (2000) *The role of cost saving and innovation in PFI projects.* (A report by Ive G., Edkins A. & Millan G.) Thomas Telford, UK.

Crosby N. & Lizieri C. (1998) *Changing lease structures - an analysis of IPD data.* RICS, London.

Dale B. (1994) *Managing Quality.* Prentice Hall International, UK.

Davenport T. & Prusak L. (1998) *Working Knowledge.* Harvard Business School Press, USA.

Deming W. (1982) *Out of Crisis.* Cambridge University Press, UK.

Department of Trade and Industry (1999a) *Our Competitive Future: building the knowledge driven economy.* White paper. London, HMSO.

Department of Trade and Industry (1999b) *Our Competitive Future: building the knowledge driven economy.* Analytic report. London: HMSO.

Edvinsson L. & Malome M. (1997) *Intellectual capital.* HarperCollins

Publishers, UK.

Egan J. (1997) Rethinking Construction. HMSO, UK.

Galbraith J. (1973) *Designing Complex Organizations.* Wokingham, Addison-Wesley, UK.

HM Treasury Private Finance Task Force (2000) *Value for money drivers in the Private Finance Initiative.* (A report by Arthur Andersen and Enterprise LSE.) UK Treasury.

Hope C. & Muhlemann A. (1997) *Service operations management.* Prentice-Hall International, London.

Latham M. (1994) Constructing the Team. HMSO, UK.

Leaman A. & Bordass, B. (1999) Productivity in buildings: the 'killer' variables, *Building Research and Information*, vol. 27, no 1, pp. 4-19.

Leonard-Barton D. (1995) *Wellsprings of Knowledge.* Harvard Business School Press, Boston, USA.

McLennan P. and Cassels S. (1998) New ways of working: freedom is not compulsory. *BIFM Conference Proceedings*, BIFM, UK.

Nonaka I. & Takeuchi H. (1995) *The Knowledge-creating company.* Oxford University Press, Oxford, UK.

Nutt B. (1993) The strategic brief, *Facilities*, vol. 11, no. 9, pp. 28-32.

Preiser W. (1995) POE and facilities management: a new diagnostic tool. *Facilities*, vol. 13. no 11, pp 306-315.

RICS (1993) *International leasing structures: a comparative study. London*: RICS.

Salisbury F. (1998) *Briefing your architect.* 2nd Ed, UK: Butterworth : Heinemann.

Schemenner R. (1995) *Service operations management.* Prentice-Hall International, London.

Stewart T. (1997) *Intellectual Capital.* Nicholas Brealey Publishing Ltd, London.

Vasconcellos J. (1999) *The war lords: measuring strategy and tactics for competitive advantage in business.* Kogan Page, UK.

Zack M. (1999) Managing Codified Knowledge, *Sloan Management Review*, vol. 40, Summer, pp 45-58

Part Five

Speculations for the Future

27   **Prospects 2000**

Facility management, as a recognised enterprise, is only twenty years old. The last ten years in particular, has witnessed a remarkable growth and diversification in FM activities world-wide. Much progress has been made in a short space of time. This has resulted in a highly competitive, maturing but somewhat confusing marketplace of FM Providers, FM Consultants, FM Contractors, FM Suppliers, FM Practitioners and 'in-house' teams (Kincaid, 1996). Now in the year 2000, an important threshold has been reached where the relevancy of FM is beginning to be recognised more widely by business and industry. Government agencies, who in the past have tended to ignore the contribution that facilities management can make, are now also beginning to recognise the importance of the FM dimension in some sectors of the economy. The UK Government's Research Selectivity Exercise in 2001 will be the first in which facility management is one of the specifically mentioned sub-areas of activity and research.

The early years of FM development have been dominated by the priorities of the office and health care sectors. Short-term corporate objectives to reduce costs, to increase competitive advantage and to enhance shareholder value have been dominant, particularly in the commercial office sector. Now the FM portfolio is becoming more balanced with diversification into leisure, retail, education, manufacturing, transport and communications, business infrastructure, the utilities and even the management of urban areas and cities. FM is also broadening its focus to address lifestyle issues and expectations, blurring the old divisions between home, work and leisure.

So FM has an interesting record of early success. The achievements made now need to be consolidated and progressed, to build a secure platform to support the next stages of development.

This concluding part of the book does not attempt to predict or forecast specific futures for FM. Instead it sets out to provide a speculative framework through which future possibilities, dilemmas, opportunities and risks may be debated. Many of the contributors to Part One of this book, looked to a future where FM will become increasingly part of core business strategy and operations. Perhaps FM will play only a modest business role here, but nonetheless the argument that it should move closer to the core of an organisation's business is compelling. Others have suggested a 'third way' resolution, combining the best of public and private sector strengths, in partnership arrangements (Montague, 1999).

The issue of public/private co-operation may be an important theme for the development of facility management, taking all stakeholder interests into account, balancing the private sector ethos (profit) with the public sector ethos (duty), and linking the private sector's experience of business and investment with the public sector's service skills and accountability. Partnership arrangements involve risk sharing. The opportunities and risks associated with the ownership and management of facilities are usually a crucial concern, since it is the relative performance of the fixed assets of a business that can give it competitive advantage or disadvantage. The two extreme positions of 'total outsourcing' as against 'fully integrated asset management' solutions, were set out in Chapters 3 and 6, respectively. The case for 'mixed' solution alternatives was described in Chapter 15. Choices across this range and particularly within partnership arrangements, give rise to three contemporary dilemmas:

- How are facility management obligations to different stakeholder interests, values, requirements and expectations to be reconciled.
- Under what conditions and for which types of organisation, might the ownership of facilities and their management be seen as either an asset or as a liability to their business for the future?
- In what circumstances and to what degree, should facility management become a more integral part of core business strategy in the future?

The view that facility management should be more sharply aligned with core business objectives, lies at the centre of Speculation 1. Here, Roger Reeves looks to a future for FM that is predominantly business driven, repositioned to provide management and service delivery at the strategic level in direct support of core business imperatives. Speculation 2 sets out a rather different and perhaps rather contrary future for FM. Phil Roberts suggests, from a public sector position, that a shift of emphasis is underway, moving from corporate business priorities towards social support provision and service quality. Here, human resource management competencies coupled to a strong ethical commitment to services provision, suggests divergence between the Business and People Trails. These two speculations serve to highlight the scale of the dilemma facing those who wish to combine the priorities of the Business and People Trails one with the other.

Speculation 1   **Repositioning Faciligy Management**

Roger Reeves, PricewaterhouseCoopers

At the start of the new millennium our experience shows that to support future business need a fundamental change in facilities management must occur with our priorities shifting in balance from tactical management to strategic direction. FM's role, as an enabler and catalyst of change, must focus on how our strategy and service delivery mechanisms for support services can assist the business to optimise its performance.

In today's business environment it is an accepted reality that work is changing, influenced greatly by employee behaviour and expectations, new work processes and patterns, and dominated by technology as business becomes increasingly dependent on information exchange, knowledge sharing, both financially and operationally. If FM is to successfully contribute in this work arena of change, and be indispensable to business in the future, it must reposition itself quickly and focus strongly on understanding and supporting core work processes as well as continuing efficiency in supporting people and activities in buildings.

To rise to the challenge of these changes in work culture, facilities managers must be ready and willing to embark on a new learning experience with the main focus on management disciplines, developing both perspective and understanding of business imperatives. With stronger management skills, reinforcement in specialist areas such as procurement and performance measurement, FM will have the best opportunity to establish a new value added relationship with its customer business.

Facilities management as an industry has emerged as one of the fastest growing sectors over the past decade. To sustain future success the FM industry needs a complementary FM profession, one which can bring to bear the analytical and business skills that give long-term benefits to our clients as well as rewards for those in the industry.

Our support offerings are innovative. FM professionals in the 21st century will be business driven, capable of operating within and on the edge of the boardroom, with the capacity to provide strategic direction and leadership for a collection of employed and independent service providers.

FM's future is promising as long as we accept that change is unavoidable, it is how we adapt to it that matters.

Speculation 2

Facility Futures in Community Services

Phil Roberts, Hertfordshire County Council

In retrospect, the Thatcher Administrations will be seen to have been a necessary, but false, start to the restructuring of community services. By 2010 many of the large scale high value contracts initiated during those years will be half way through their term. Some providers will have been unable to manage the risks, and restructured deals which increase the cost to the taxpayer will have been agreed, Some public clients will have found that new technology fundamentally changed the way they provide services. They will also have been forced into expensive re negotiation of long term agreements.

The Blair Administration's Best Value initiative stimulated a new generation of partnering and outsourcing agreements. These second generation agreements focused on service quality and citizen involvement rather than large scale corporately managed investment. Their purpose was to achieve educational outcomes, not maintain school buildings, and improve public health, not manage hospitals. It also became apparent that if community services were to be designed with the individual citizen in mind, existing local government institutions needed radical reform. As a result many existing community service organisations such as educational authorities and health trusts were wound up. They were replaced by regional quangos tasked to oversee the provision of integrated community services and to set standards for community involvement within towns and cities. Actively managed regional land and property banks were created from the assets of the old institutions to finance the programme of change. This new framework provided the opportunities for the new breed of service provider.

These second generation providers used new technology to involve stakeholders in service design and to report outcomes to the community. They specialised in total service quality, taking responsibility for all aspects of service delivery including staff, logistics and facilities. Rather than originating from the construction and property industry, they emerged from specialist service delivery organisations and IT, business support, retail and voluntary sectors. Their core competencies lay in human resource management and the effective management of knowledge. They placed a new emphasis on service analysis and design which linked the workplace to customer-led service outcomes. Long term inflexible contracts were replaced with shorter term renewable partnering agreements with open book assessments of shared risk. Above all, the second generation providers worked within a strong ethical framework committed to improving the value they provided to the community through self managed service improvement and open dialogue with community stockholders.

Looking back from the perspective of 2010, the old arguments about core and non-core, outsourcing and insourcing, and the meaning and definition of facilities management, will be seen to have been only stepping stones to radical new strategies for the provision of community services.

The second part of this book speculated that the HR 'People Trail' might be the most radical and dynamic route to the future of FM. Social and cultural change, rather than technological change, might be about to lead the next stages of the revolution relating to work, the workplace lifestyle and leisure. The increased choice and diversity of work styles, work times, workplaces and work locations, gives rise to major dilemmas for organisations and their HR management. The potential impacts, both positive and negative, of these developments on facility management policy, purpose and operation, raises further dilemmas about how work will need to be supported in the future.

Retaining key staff is now a major concern of organisations world-wide, particularly in the USA. Organisations are looking for ways to harmonise the working and personal lives of their employees to improve the quality of their work and life. With flexible working the facility manager has a central role to play. The 'real' complexities of the changes that are underway need to be addressed urgently, with FM learning how to support the 'real' patterns of work of 'real' people in 'real' time, rather than the management of static workplaces and dedicated workstations alone. Key dilemmas include:

- How to develop a secure organisational culture and management structure, with appropriate work processes and procedures for effective flexible working, resolving the new tensions between individual freedom and organisational control?
- How to determine which functions and work operations are suitable for flexible working, which are not?
- How to decide which types of employee, when properly trained, might be suitable for various degrees of flexible working, which not, taking into account their age, gender, role, experience, the stage in their family life-cycle, their motivation, commitment and other relevant circumstances?
- How to develop diverse and flexible forms of employment contract for new ways of working, deciding on the degree to which future employment packages should include facility, technology and service support levels of agreement and other inducements, when a significant proportions of off-site and home working is intended?
- How to combine 'new' and 'old' ways of working with an appropriate 'mix' of working modes across the four domains of work shown in Figures 18 and 19 on pages 74 and 75, managing adjustments over time as business pressures change?
- How to predict the needs of individuals and teams when working at varied and remote locations, and to quantify their demands, together with those of the business generally, for facility and services support under flexible working arrangements?
- How to quantify and deliver appropriate facilities and logistical support to teams and individuals, where and when it is required, but with efficient levels of utilisation?
- How to monitor and measure the appropriateness and effectiveness of flexible working support systems, to ensure FM accountability for the management of a diverse and dispersed set of facilities and support services?

While there are a few embryonic 'good' practice exemplars here, no generic procedures for facing these dilemmas yet exist. New ways of working will place much reliance on the self-motivation, self-discipline and self-management of individuals and working teams, with a degree of self-determination of work time and workplace. This will need to be backed up by more complex and responsive facility management arrangements, involving some delegation of basic FM functions to project teams. Active employee participation in parts of the FM process of support will be essential.

Table 23 summarises some of the opportunities, benefits, potential risks and constraints, that new ways of working might entail. The opportunities and risks are allocated to four of the main stakeholder groups, corresponding to the concerns of business management, human resource management (the employer), the employee and the facilities management team. Stakeholder priorities are quite different both between employer and employee perspectives, and between individual and organisational levels of concern. It would seem that the majority of opportunities offered by the new ways of working tend to promise short-term benefits in contrast to the potential risks, which appear to threaten stakeholder interests over the longer-term. So the challenge here will be to develop secure management structures and procedures that can consolidate and preserve the short-term benefits, while containing and avoiding the long-term risks.

Table 23   Opportunities and Risks on the People Trail.

Opportunities	focus	Potential Risks
Business Flexibility Reduced Cost (Space) Increased Productivity Close to Customer	**Business**	Company Culture Organisation Loyalty Business Security Corporate Visibility
Flexible Employment Workforce Choice Responsive HR Deployment	**Employer**	Legal Minefield Staff Moonlighting Intellectual Property Loss
Individual Freedom Working Flexibility Self Pacing and Choice Job Variety Working Privacy Job Satisfaction Lifestyle Flexibility	**Employee**	Work Isolation Work Motivation Career Development Company Disaffection Self Esteem Family Pressures Job Security
Performance Driven Work Output Criteria Work Team Participation Services Delegation FM Downsizing	**Facilities Management**	Dispersed Management Performance Measurement Services Accountability Collaborative Working FM Visibility and Value
short-term benefits		**long-term risks**

Speculation 3 considers how these changing opportunities along the HR trail might impact on the built environment of the future, particularly on 'distance' and 'location'. Stephen Brookhouse suggests that they will result in fundamental changes to the ways in which facility environments, both physical and virtual, are procured, located, designed, used, managed and adapted, linking the concerns of designers and managers, one with the other. Speculation 4 develops this theme further. Barry Varcoe looks at the dilemma that is faced when attempting to reconcile today's speed of change in demand for property, with the slow rate of possible physical change on the supply side. Greater fluidity in tenure type and duration, fluidity in capital asset management, and more fluidity of use over time, is predicted. A future that integrates all of the functions of facilities management, facility design, project management and occupancy planning, is foreseen. This integration could result in an infrastructure resource management platform, around which the property industry could begin to restructure itself.

Speculation 3

Managing Design after the Death of Distance

Stephen Brookhouse, Design 5

Frances Cairncross, in her comprehensive survey of the effect of the communications revolution on our everyday lives, the Death of Distance, predicted that thirty key characteristics will define the way we live and work in the immediate future (Cairncross, 1998). A small but significant number will directly affect the future of the built environment and the way we procure, design and manage our existing and future facilities. The remainder will no doubt have an indirect effect on our everyday and professional lives.

Cairncross wisely differentiates between new and emerging technologies. The basic technology and potential for electronic communication are already in place making predictions about the future easier and less speculative. Indeed, one of the key characteristics, the Inversion of Home and Office, is already taking place at the margins of the working environment. Growing numbers of professionals in a wide range of organisations are happy to allow the line between home and office to become blurred by using the office to celebrate the social rather than production aspects of work. In turn both the 'office' and the 'home' as building types and icons will begin to adapt on a wider scale to these changing requirements. The re-birth of the city as a place for living and entertainment rather than commerce identifies a trend that is already well-established and which seems to be driven more by the shift away from manufacturing and storage of goods to the provision of high value services.

However, the 'fate of Location' as a key business driver coupled with the 'irrelevance of size' and the continued recognition of 'people as the ultimate scarce resource', rather than land, will have deep and lasting effects on the way we procure, locate, design and manage new buildings. They will also begin to affect the way we evaluate, adapt, demolish and renew the built environment.

The strategy for successfully managing the future built environment is likely to bring together and develop the current incremental improvements in design management and facilities management.

Design and facilities professionals will continue to develop methods that allow them to identify, understand and respond to users. As a result, the briefing and design management process, by bringing together the business case and identifying the links with user requirements, should emerge as a critical activity carried out by a multi-disciplinary team of professionals.

Dynamic briefing using innovative management methods and a 'self-learning' approach resulting in the continuous optimisation of value and the reduction of procurement times will allow a more fluid and effective approach to decision-making.

As the existing physical stock and categories of building become obsolete faster, a systematic approach to the evaluation of the built environment in use will lead to design and facilities managers feeding these lessons forward using objective measures. The current fragmented and sometimes anecdotal evidence about building performance will be built into a body of knowledge that will allow the design and facilities team to concentrate on the actual rather than the perceived success factors. Construction organisations will continue to use innovation and standardisation of processes as well as products to reduce construction costs and cost in use.

On the surface, the built environment responds far more slowly to technological change than other areas of capital investment. Indeed certain locations and building types benefit from the dramatic growth in increased speed in communications. Getting closer to major and leading edge organisations with the objective of improving value by satisfying business needs seems to be a sound strategy, particularly when technological change has only a limited impact on the technology of construction. This is a less secure approach when one begins to distinguish between technological change that supports the organisation and disruptive technologies that overtake and undermine its business performance. For example, at a recent conference a major American bank predicted that any financial institution not actively developing e-commerce as its major line of business is unlikely to survive beyond 2005. Refining existing systems is unlikely to be adequate in the face of e-commerce's disruptive effect. As a consequence the size and location of related buildings will diminish in importance.

Designers and facilities managers will need to innovate 'outside the box', predict and plan alternative scenarios, remove professional barriers and take risks. They will need to lead rather than respond to innovation. In a world where the virtual and built environments rub shoulders design and facilities management will not only learn to find new uses for redundant building shells but also design and manage environments that may never exist.

Speculation 4   Fluidity

Barry Varcoe, Johnson Controls Inc.

It is now a commonly accepted observation that there is a growing conflict between the indigenous speed of thought, action and change of commerce, and the property industry's ability to adequately respond. It is now evident that the pressure that this gap causes will in turn lead to significant change throughout the property industry. Several examples are apparent already; more are to be realised at a general level. This speculation considers some of these changes in the context of improving the property industry's 'fluidity' at four levels.

Fluidity of Availability

The pressure for increased fluidity of tenure has already lead many to broadly cluster properties within their portfolios with respect to their likely length of occupancy. Long term 'core' accommodation is owned or held on a long lease, medium term is on shorter leases (1 to 5 years, or longer given good break provisions) and the short term and/or remote requirement is increasingly being met by the burgeoning serviced-office sector. Increased sophistication will come as real estate groups consider tenure duration at a portfolio level with respect to at least two new perspectives - commitment and control (Deeble, 1999). By using probability modelling they should be able to achieve a much improved understanding of both the minimum amount of space they should be committed to for different time periods, and similarly the maximum amount they should have control over (i.e. a right to occupy, but not an obligation).

Fluidity of Capital

The globalisation of capital markets has made the availability of capital a commodity. Its affordability is another issue, however, and many organisations are looking much more closely at the significant amounts that are often tied up in its property. Many innovative financial vehicles are appearing that offer the potential to unlock and redistribute this capital, based on operating lease type arrangements, but these tend to be focused more at an asset by asset level, particularly in the commercial sector. Portfolio level operating leases is a logical further step for some, but there are inherent consequential dangers regarding duration commitment and unit cost. More generally it is to be hoped that investment analysis within real estate becomes consistently more sophisticated, and that multiple scenarios and risk be considered before decisions are made.

At another level there will also be an increasing consideration of the trade-offs to be had between property and IT/communications investment, with the probable consequence being more for IT and communications allowing a consequential less for property (on the basis that better 'wired' office can be used more intensively, meaning that less of it is needed for a given amount of demand).

Fluidity of Use

Organisations have been looking at and taking advantage of 'time' as a resource for a number of years, typically through a range of non-conventional workplace concepts sometimes collectively known as 'alternative officing'. The next step forward could well be to take advantage of the improved efficiency and effectiveness to be gained by 'change in time' - allowing work settings to be re configured in little time for little or no cost. Static design and static layout will become a thing of the past, and accommodating the needs of the minute, hour and day will be 'business as usual'.

Fluidity of Management Approach

The professional disciplines still predominate, and their influence has caused too many functional 'silos'. The information revolution dictates that decision-makers today must have access to a complete perspective of relevant issues, not a detached and incomplete patchwork. The first step will be an integration of the all too often separate disciplines of facilities management, design, occupancy planning, project management, property management and transactions/agency, at least form an information perspective, to create a true asset by asset perspective of how the workplace is meeting all the various needs of its various stakeholders. Beyond that there should then be aggregation of the assets to achieve a true portfolio perspective, thereby enabling fluidity and responsiveness to the organisation as a whole.

If the property industry fails to move quickly enough to change this, almost certainly others will do it for them. The consolidation of enterprise resource planning and customer relationship management systems into the new breed of 'enterprise resource management' systems will push this if nothing else does. They will embrace infrastructure resource management, which will not only force the cohesive management of all aspects of 'property', but will place it firmly in the context of the overall 'production platform' of the organisation (including IT, finance, HR, etc.).

Fluidity of Existence?

Without doubt the property industry is beginning to change fundamentally, and there is far more to come. It is in transition, the unabating pace of change in society and business at large forcing it to face up to its structural deficiencies. Indeed, some of its newest components, such as FM, may in time be seen to have been merely a transitory phenomenon - a hitch-hike on the road to a new horizon.

Some of these signs of change within the property industry were identified and considered in Part Three of this book. However, the inertia of the Property Trail makes it the most difficult route on which to change direction to a better future, certainly during the first decade of this new millennium. Here, making good use of the latent 'flexibilities' of property has been a continuing theme. Today's demand seems to be for flexible facilities, flexible leases, flexible space, flexible services, with flexible legislation to support flexible business throughout the facility life-cycle. The potential value of flexibility is recognised widely by those in finance, planning, design, business and management. So flexibility has many facets. But it also comes at a cost.

"If the demand side wants to have flexible short term leases and avoid the risks of property investment then they should not be surprised that the price will be high, because someone else will need to be paid for taking that risk".

(Professor Hans de Jonge, Delft University of Technology, 1999)

For more than thirty years there have been 'good intentions' for achieving greater flexibility throughout the property life-cycle. Concepts of life-cycle briefing, life-cycle design, life-cycle costing, life-cycle use, life-cycle management, life-cycle adaptation and life-cycle demolition, have been discussed endlessly, but few reliable measures and methods for the practical implementation of 'flexibility' on the ground, have been developed. Attention to the 'real' life-cycle issues of demand is now being sharpened through the introduction of PFI/PPP initiatives, as described in Part One of this book.

"PFI is causing a change in the way that we look at buildings, a change from the 21 year repairing lease, the 'let and forget' approach property, to an environment in which property has to be serviced and the demands of the occupant need to be met more flexibly ... PFI is also an agent for change in making contractors into investors. It is forcing contractors to align their expectations of reward with the economic life of the asset and not just the construction contract duration"
(Adrian Montague, Chief Executive, Treasury Task Force, 1999)

Developments such as these are stimulating change within the property industry. The Property Trail still starts out from a position that is 'supply-led', where the interests of the property investor and the property professions, tend to dominate the interests of the property customer and consumer. However, there are encouraging signs that attitudes are changing and that some in the industry are developing a more sophisticated and well balanced position, responding to the strategic 'demand side' requirements of organisations and the operational needs of their employees. From a management perspective, FM offers a direct opportunity to link the 'post-occupancy' priorities of property use and operation back to the 'pre-occupancy' concerns of facility planning, briefing, design and construction. It promises to create the means for an effective dialogue between those who use the built environment and those who produce it.

But what of the design perspective? Are architects addressing the design opportunities and dilemmas of new ways of working and the novel demands that these are generating? Yes, some are, as evidenced by British Airways' business centre at Waterside (Facilities Management Excellence, 1998). The RIBA is promoting a clearer client and customer focus for designers, but as their publicity material shows, the profession still needs to give greater attention to the management of the design product once in use (RIBA, 2000). Rather radical questions arise in relation to all of the key areas of design decision, linking pre and post design concerns. Take just one example, the choice of 'appropriate technology'. First, just how technologically sophisticated do future facilities really need to be, given current developments in cordless technology ? Second, what will be the probable shelf-life of any 'design embedded' technology and how easily could it be upgraded in the future? Third, how robust will the selected level of technology be in facing the probable patterns of changing demand, the likely rate of technological obsolescence, and future market conditions? Each of these questions relate to the risks of 'over' or 'under' specifying the level of technology to be incorporated within any specific property product. This is but one of the designer's dilemmas, there are many more. The major dilemmas for the future will certainly include:

- How to design, develop and manage property to provide strategic and operational flexibility to the property customer, to support their dynamic patterns of demand throughout the facility life-cycle, both the demands that can be anticipated and those that cannot?
- How to determine an appropriate technological, utility and environmental specification for a given property to support its prime function and alternative functions, within its market niche at a given location?
- How to regulate property development, building design and facility use in a future where the concept of 'functional types' becomes irrelevant, invalid and obsolete.
- How to bridge the gap between the initial suppliers and the ultimate users of property, linking the 'post-occupancy' priorities of business organisations, facilities managers and end-users for the effective 'life-cycle' operation of facilities and services, back to inform the 'pre-occupancy' concerns and skills of those involved in facility planning, briefing, design and construction?
- How to encourage the development of innovative property products that anticipate rather than follow market demands, with higher utility values and robustness so that they are more manageable, adaptable and sustainable for the future?

Developments along the Property Trail suggest new opportunities and roles for facility management, both in relation to the formation, operation and regulation of the built environment. Many of the authors in this book have identified new social and environmental responsibilities for the FM function. In Speculation 5, David Baldry sees FM becoming the natural 'managing agent' for the implementation of much of the environmental agenda, particularly in regard to policies for more sustainable property futures. FM could become the primary vehicle to encourage business organisations to adopt sustainable policies as part of their normal business plans and operations, adapting measures that may be taken to improve sustainability through the modification, renewal and adaptation of facilities. Pivotal roles of this kind would generate a new set of ethical expectations and duties for FM, linking contemporary environmental requirements, with sustainable facility management practices, in support of sustainable corporate business goals and operations. Finally in Speculation 6, Partrick O'Sullivan suggests that the means of building and environmental regulation may need to change radically. He suggests that the 'regulation through design' may need to give way to a future that is much more reliant on 'regulation through operation', again with the facilities manager in a leading role.

Speculation 5   **FM as Agent of Environmental Responsibility**

David Baldry, University of Salford

The creation of a building invariably results in extensive, often profligate, consumption of resources and materials, with substantial and often irreversible impacts upon landscape, amenity and ecosystems. Similarly the management and operation of a completed built facility demands significant resource usage, particularly that of energy, whilst seeking to create healthy, safe and fit-for-purpose physical environments to meet social and economic needs and the satisfaction of corporate operational objectives. This speculation considers the prospect that effective facility management can be, and will probably need to be, the agent for the long term, durable and sustainable operation of built environments in the future.

There is a growing awareness that industrial and business practices will have the most significant impact upon the future well being and sustainability of natural systems. Sustainable development "recognises that economic growth and environmental protection are inextricably linked, and that the quality of present and future life rests on meeting human basic needs without destroying the environment on which all life depends" (Schmidheiny et al., 1992). The conventional role of facility management is to act for the client/occupier in such a manner as to seek to optimise performance criteria in respect of technical performance and operational satisfaction. This pursuit of optimisation inevitably results in the sub-optimisation of certain outcomes which may include environmental impact. However the success criteria for evaluating operational performance is increasingly extending beyond compliance with technical standards to embrace some degree of social responsibility and ethical behaviour. Sustainable facility management practices are capable of being delivered where a condition of mutual benefit can be created, that is where a sustainable approach can be recognised as coincident with sound business practice whilst still being driven by market incentives and stakeholder pressures (Hutchinson & Hutchinson,1996).

The discipline of facility management offers the potential to exercise the greatest influence to achieve the formation and operation of built environments in ways which recognise and satisfy both operational and environmental needs. This influence will arise from:

- The increasing awareness amongst facility users and managers of the resource implications of the facility life-cycle.
- The ascendancy of the facility function in corporate decision-making with a predisposition to contain space formation and resource consumption.
- The growing acceptance of the leadership role of the facility function in major project development and realisation.
- The increasing national and international expectations for resource efficient operation and commercial activities and the raising of ethical sensi-

tivities associated with the exercise of commercial responsibilities.

The view of the economist Milton Friedman (1984) that "good business is good ethics" is persuasive and engaging. The inverse of that statement has historically not been widely found to be true but it is suggested here that the facility management discipline will be increasingly required to be one of the primary agents for the achievement of corporate objectives whilst upholding standards of ethical performance including, specifically, those of environmental responsibility.

In order to exercise this pivotal role an increasingly proactive and assertive facility management discipline will need to emerge, one which has more effectively coalesced its activities into the primary business of the host organisation, rather than remaining as a support activity responding to a prescribed set of expectations and duties.

Major strategic steps will need to be taken to integrate the facility function into business processes in order to meet raised ethical expectations of customers, legislators, regulators, and other significant stakeholder interests. Facility management will be required to be the catalyst for corporate activities which can achieve coincidence between successful business practices and goal achievement, and the realisation of environmental and ethical modes of operation.

The achievement of the above will depend upon the raising of facility management awareness and expertise from the level of the operation of technical and managerial systems to that of whole business planning and delivery within a more demanding framework of social obligations and environmental responsibility. This will be brought about by the culture of a breadth of vision which can find common ground between sustainable facility practices and corporate goal achievement, as noted by the CEO of a major industrial organisation (Margretta, 1997): "far from being a soft issue grounded in emotion or ethics, sustainable development involves cold rational business logic". This will be one of the major and most important challenges facing the professional management of facilities.

Speculation 6 — Regulation through Operation or Design?

Patrick O'Sullivan, University College London

Buildings are not merely locations where people meet, work and live with reasonable levels of efficiency and health, they also make a significant contribution to our social and aesthetic habitat. In the past this habitat has been policed by land-use planning and regulated through building design. Today the dynamics of business and the rapidly changing patterns of use, work and leisure are challenging our dependence on static regulatory systems of this kind.

The emergence of facilities management offers unique prospect of regulation through the operation and management of the built environment, where appropriate. As such, the opportunities for facility management lie not in the use, operation and maintenance of facilities alone but in the potential to offer a comprehensive 'life-cycle' service to building occupants so that they can perform their chosen business, educational, health, domestic or other activities as efficiently, effectively and safely as possible. Facilities managers will need to have a continuous dialogue with end users, asking:

- What can we do to help you operate more effectively, more productively, more healthily and happily?
- What are your expectations for facilities and support services, and what affordable standards do you require?
- How can we help to reduce the stresses on you in performing your operations and work tasks?

The potential remit for facility management is excitingly wide-ranging. It could involve the employment of secretaries and e-mail clerks, the maintenance of personal communication and computer systems, work and leisure travel arrangements and so on, all of equal importance to the operation of the heating and ventilating plant! It could also include advice on how to operate facilities during major refurbishments and the commissioning of new facilities, with criteria for tomorrow's handover for 'operation' in contrast to today's handover for 'maintenance'.

In the UK, and perhaps in many countries around the world, we have the majority of buildings we will ever need. All of the modern tendencies are reducing the space requirements per person within buildings. The emphasis of the future will therefore be on the resolution of space issues within the existing building stock rather than acquiring more space per se. We must anticipate that the more efficient use of buildings, the higher the utilisation of building space, then the higher the energy consumption per square metre, the higher the CO_2 emission, and the higher the operational, refitting, maintenance and adaptation costs in the future.

In such a situation we should perhaps expect that our regulatory base will reflect this change, moving from a predominantly 'design', to a 'design and operate' mode. If this were to occur, then the importance of commissioning would increase. The distinction between completion for 'occupation' and completion for 'maintenance' would sharpen. A facility's spatial logic will need to be consistent with a security logic, a safety logic, an energy logic, a cleaning logic, a maintenance logic, an adaptation logic, and so on. The need to explain, determine and improve the sustainability 'quotient' of the building and its site may perhaps become increasingly mandatory, with all that this may implies in terms of a climate levy, capital gains taxes and 'sustainability' allowances.

Regulation through operation will need a strategic and policy dimension. In such a situation, the people responsible for the operation mode of our buildings and their attendant regulations will have a chance to lead this part of our future and themselves become a central part of 'the new facilities provision'.

They will be the new team leaders. A role that will require the ability for self certification with the attendant education, training and quality control that this implies. What an opportunity.

References

Cairncross F. (1998) *The Death of Distance*, Orion, London, 1998.

Deebl, K. (1999) *Financing Corporate Real Estate: a raw materials procurement approach*. The Corporate Real Estate Portfolio Alliance, April.

Facilities Management Excellence (1998) *The Global Village: case study of the Waterside develoment*. March, 1998.

Friedman M. (1984) The Social Responsibility of Business is to increase its Profits. In: *Business Ethics* (1990) (Ed. by W. Hoffman & J. Moore.) McGraw Hill, New York.

Hans de Jonge, Conference Transcript from "Facility Performance of CRE/FM Units", *Conference in Property and Facility Management*, UCL, London, June 1999.

Hutchinson A. & Hutchinson F. (1996) Environmental Business Management: Sustainable Development in the New Millennium, McGraw Hill (International) UK.

Kincaid D.(1996) An Overview of Facilities Management Development, Paper presented at the RICS *Conference: Switching to Facilities Management*, London, 16 May, 1996.

Margretta J. (1997) Growth through Global Sustainability, *Harvard Business Review*, Jan-Feb., pp. 79-88.

Montague A. Conference Transcript from "The Private Finance Initiative ; directions for the future". *Conference: Futures in Property and Facility Management*, UCL, London, June 1999.

Royal Institute of British Architects, *Architects and the Changing Construction Industry*, RIBA PRACTICE, 2000.

Schmidheiny S. & The Business Council for Sustainable Development [BCSD] (1992) *Changing Course: A Global Business Perspective on Development and the Environment*, BCSD and MIT Press, Cambridge USA.

28 The Platform for FM

At the start of this new millennium, the 'knowledge business' and 'knowledge management' dominate the future agenda. Here, the facilities management service is vulnerable. The Knowledge Trail, as explored in Part Four of this book, is the most primitive and undeveloped of the four FM trails considered. It starts from a position that may be characterised as information 'rich', data 'unreliable' and knowledge 'poor'. Technical information in the FM field tends to be 'strong' but management knowledge is 'weak'. Furthermore, FM is a field like many others, whose foundations are being shaken by the new technologies. The task ahead is to transform 'information', of which there is much, into 'knowledge', of which there is little, for practical 'application' across the whole range of FM sectors, services and circumstances making full use of all available technology and exploiting further opportunities as they arise. The initial step in this process will be to broady converge on the way ahead, with a sharper definition of the core territory of FM to:

- Clarify the distinctive features of facilities management knowledge, clearly defining the unique FM functions that are not part of the fields covered by business management or the property professions.
- Develop management concepts and expertise in these unique areas, focused on the specific roles of FM in managing resources, environment and services to provide logistic support to the operations of organisations.
- Adapt, test and apply relevant management concepts and proven technical expertise that can be modified to directly support the key areas of FM practice.
- Build an expert knowledge structure and data base with methods and techniques to support these functions, roles and practices of FM.

The chapters of this book and the trails that they have followed, provide clear evidence of the extensive breadth and diversity of FM concerns. So to those that suggest that FM is an emerging and new disciplinary area of development, the answer must be 'not so'. The facility management remit is far too wide and complex to be considered as a unitary 'field of study' or as a new 'discipline' in its own right. It covers what is a natural cross-disciplinary territory of service and opportunity (Nutt, 1991). Two questions arise. First, how to consolidate and create an appropriate FM knowledge structure and data base across this wide-ranging and diverse field? Second, how to realise its value in application?

In response to the first question, many have argued for an integrated approach (Kincaid, 1994). On the supply-side, it has been suggested that integration will give coherence to the partial and fragmented areas of information, skill and knowledge that currently constitutes the FM field. On the demand-side it is argued that an integrated FM service is what the customer requires and that an integrated approach will be beneficial and better able support business needs.

This usually unchallenged assumption that 'integration is good', may need to be questioned when developing the FM knowledge structure. Fully integrated systems involve risks as well as benefits. They can be inflexible, difficult to change and adjust. The issue of flexibility is of equal importantance for knowledge management as for the physical, functional and financial dimensions of flexibility that were discussed in Part Three. The need now is to move beyond obsessions with bench marking, static standards and areas of competency, to create a flexible knowledge platform that encourages and supports diversity and variety. The FM knowledge resource must be capable of continuous adjustment and adaptation to meet the ever changing needs of organisations and the huge variety of requirements organisation to organisation and sector to sector. This suggests the development of 'smart knowledge' systems, with the ability to re-differentiate, re-configure and re-distribute knowledge rapidly, in response to different application circumstances and requirements. So re-differentiation rather than integration may be the way ahead.

'Smart knowledge' is generic, versatile, robust and quick. FM knowledge will be 'generic' when it provides a fundamental understanding of recurrent issues in the field, built around a framework of 'practical theory' to support further development over time. FM knowledge will become 'versatile' when it is widely relevant, applicable and adaptable to all sectors, and is potentially useful across the full range of circumstances and conditions that it can encounter, remembering that 'situation specific' rather than 'generic' solutions are usually required. FM knowledge will become 'robust' when its life expectancy lengthens, and its lines of application become keener, secure and sufficient for its ultimate practical purposes. Finally, FM knowledge will become 'quick' when it learns to make full use of all available technology and mutates to a form that can exploit the opportunities of the internet and associated technologies.

Collaborative work to development a cross-disciplinary knowledge base has begun and is ongoing, linking the areas of available expertise with the areas of potential application (Nutt, 1998). Many of the building blocks are in place already. There are some forty or more 'expert' FM groups, units and centres in universities, research institutes, international consultancy firms and advanced practice units around the world. Each could contribute to the development of FM's intellectual property base, of course with different strengths and biases. Over the next few years these initiatives should be consolidated and applied urgently, developing secure management and decision techniques, based on sound criteria and accountable expertise. But is technology running ahead of knowledge ?

Speculation 7 considers the potential impact of internet technology on FM and the businesses that it supports. Richard Byatt speculates on the ways in which 'real time' information and 'speed' might transform the practices of property and facilities management in the 'near' future. With this technology, many FM operations and much of their regulation and control, could be managed electronically off-site. In Speculation 8, John George spells out a future in which property and facilities could, in the main, be operated remotely and automatically via the internet. This could relieve facilities managers from many of the day to day practical operations, allowing them to concentrate on their strategic role in the management of sustainable and effective work environments.

Speculation 7

e-FM

Richard Byatt, i-FM.net

The facilities manager is on holiday (and on her second cocktail) when the call comes. The message, from the VP Global Operations, is succinct: "We've just won the licence for broadband communications in the Czech Republic, we need to put a team in there quick. I want space and support in Prague for ten people by Thursday."

The FM logs onto the Corporation's intranet, calls up the global property pages and runs a search for Prague. A recently acquired subsidiary has offices there. "Probably still swimming in space," she thinks and checks the plans. With a bit of re-stacking there is space on the third floor – she fires off an e-mail to the team leaders and then posts a requirement for support to the Office Services R Us auction site. While waiting for the bids, she checks the Czech project web site, arranges the re-routing of all e-communications to the Prague team and reserves accommodation through the Corporate deal with Biztravel.com. The ice in her drink is just beginning to melt.

The internet tends to polarise views. There are those who believe it is changing the rules of business. Others, the majority probably, say it is just another way of doing business as usual. I believe we have seen enough in recent years to predict that internet technology will turn many conventional business models on their head – and that's before access via mobile phone and digital television turns the net into a truly mass medium.

Everything is zeroes and ones. The digital age requires a new way of thinking about information, all information. Whether it's a text document, a spreadsheet, a drawing, a photograph, a phone call – it all ends up as a series of zeroes and ones. Once you start thinking about information in this way, huge possibilities open up and facilities managers deal in information.

The application of information technology to facilities management has been characterised by over promising and under performing. FM began by borrowing IT from other specialisms – architecture, space planning, engineering maintenance, human resources. Gradually, developers brought out specialised applications and CAD became CAFM became CIFM became infrastructure solutions.

Now, internet technologies such as Extensible Markup Language or XML offer a common language, an intuitive interface and a set of tools to be used by anyone. For facilities management, the textbook cross-discipline activity, the internet provides the means to combine the big picture with the essential detail, to give selective or open access as appropriate and to devolve information management.

The early adopters of internet technology in FM used it to re-publish information, to promote themselves and their services. They have since moved on and are beginning to integrate the internet into their activities. This trend, accelerated by the demands of customers and clients for 'real time' information, will see more project web sites, on-line concierge services, knowledge banks, mobile maintenance systems and extranets.

Some entrepreneurs see the internet changing conventional accounting. Individuals will become cost centres, charging not only their time but all their resources and support to projects and free to buy products and services in the wider market, beyond corporate purchasing arrangements.

The internet is changing work and making possible new ways of supporting workers but it is also giving people ideas for new types of business. In the construction industry a number of portal sites, on-line trading communities, are already fighting for dominance in the US and Europe.

In property and facilities management, the portal initiatives are coming from real estate firms, heavily backed FM providers and technology companies. By the time this book is published we are likely to see the first ventures, selling office products and 'commodity services' to a whole new market which it has now become cost-effective to serve.

The crunch questions for these e-FM businesses are how far can they move along the product-service continuum, how much value can they add and what margin can they make? They may have to cannibalise their existing businesses to make the transition to the new economy. In that they will be no different to their customers.

One word of caution in all this. The defining characteristic of e-business is speed. Simply the ability to do things more quickly, not necessarily better but certainly faster. Although the internet holds out the promise of the perfect market – all buyers and sellers equally well informed – the reality will be very different. The distinctions between information, data and knowledge become even more blurred when you're in a hurry.

Speculation 8   ## Remote FM

John George, Barbour Index

To speculate for the future is to engage in theoretical reasoning and to consider abstruse and uncertain matters. In reviewing the future of FM there are some certainties. The function will exist and be higher in the hierarchy of business management responsibilities. However, the job of facilities manager as it is known today may not exist; if it does it will be very different.

In the 21st century, indeed during the first quarter, buildings and the processes they house will be managed off-site from a central control point. Some fire and security systems are managed in this way today, but similar techniques will be used to assess the state and performance of buildings and the functions they perform.

The technology is now available and will be applied to measuring and reviewing on a programmed periodic cycle. All elements of a building and the facilities it provides - structural stability, movement, heat loss and gain, drainage systems, humidity (wet and dry rot) - will be measured and reviewed hourly, daily or weekly, as appropriate.

Sensors will be fitted, linked to electronic systems to a control point with pre-determined parameters for performance so that any signal outside the norm will generate a warning.

The benefit of these systems - which will be incorporated into new buildings as standard and retro-fitted to the majority of others - is that sensors may be placed in locations where performance is critical and where access is often difficult. The main advantage is warning of weakness or non-performance of building elements before a fault becomes manifest in structural failures, or rot and damp appear, and serious safety and operational issues arise.

One simple example: in storms, drain and gutter water flow can be checked frequently and any design, maintenance or clearing problem dealt with at once. The placing of sensors in roof spaces, cellars and storage areas will allow conditions to be monitored and provide records from which performance trends and early warning of problems can be gained.

Generally, managers are aware of problems only when they become serious and manifest in cracks, rot etc., often requiring expensive and disruptive structural restoration.

As a consequence of these systems being available, building owners, and particularly those designing for their own use, will require sensors to be incorporated in initial designs. They will have the benefit of lower costs of management and preventive maintenance on an as needed basis with the economies that will result. There will thus be an even greater need for those experienced in managing and operating buildings - facilities managers - to play a key role in the design process.

Thus, buildings will be managed by an integrated system of electronic controls recording performance through sensors strategically located to measure the whole range of key data to ensure efficient, economic operation and maintenance. Systems may be interrogated through dedicated intranet systems or via the Internet. The facilities manager, whilst perhaps located remotely, will be able to maintain a closer and more rigorous control.

Perhaps sensitive sensing rather than remote FM

Speculations of this kind, and there are many of them, serve to emphasise the extreme levels of uncertainty that are faced today, both about the risks and the opportunities that the ICT revolution is generating. While ICT 'lead times' makes future developments in technology reasonably predictable over the short to medium term, the actual course and extent of its application to business and work practices and its impact on society generally, tends to be highly unpredictable, even for the next few years ahead.

> "Working at home and operating on the internet may be part of the answer, but it certainly in many ways is not the only part, but one must foresee a great many changes. I think that the whole history of scientific revolution shows that a pretty small change in something can lead to enormous repercussions upon all our lives".
> (Sir Crispin Tickell, Chancellor, University of Kent, 1999)

The 'real time' electronic business environment has the capacity to collect, store and use vast amounts of information that were previously uneconomical to manage. It is also creating entirely new areas of service provision. The creation and use of knowledge within organisations is now seen as the new source of competitive advantage in business. It may be no different for facility management. The accelerating growth in e-commerce developments through web-based business-to-business activity is, on the one hand overturning conventional economics based on the economies of scale, yet on the other hand it is demonstrating the continuing usefulness of many conventional procedures, such as transaction cost analysis for example. While the former continues to captivate the imagination of the financial markets, it is the latter within which the real productivity gains are being made. Within some FM areas of service procurement, transaction cost reductions as high as 80-90% are being achieved (Economist, 2000). Many electronic banking operations also report savings in this range. What seems clear, once the 'hype' is dispersed, is that ICT should allow businesses and some of their associated FM activities, be to conducted in new and novel ways at an enormous benefit to their customers. Some are more sceptical:

> "The rhetoric of the last five years has contributed to the alienation that a lot of customers feel, but also people in business, at the exaggerated claims made for information technology. So professionals are beginning to question whether or not the productivity gains offered by ICT are real, with citizens in general asking what is happening to their quality of life as these systems penetrate every aspect of our lives and surround us".
> (John Thackara, Director of the Netherlands Design Institute, 1999.)

It will perhaps take some years to demonstrate the variety of productivity gains that can result from the application of this technology to FM. It already seems clear however, that ICT applications hold particular promise for those that are engaged in customer focused services, since they provide the means of obtaining, holding and analysing information on the profile and likely requirements of specific customers. The ability to unlock specific end-customer data offers a valuable opportunity to those wishing to provide a personal service. For example, banks using customer information to cut out irrelevant offers and tailor their products to the individual, are enjoying increased revenues and customer loyalty. There are similar opportunities in the FM field, particularly in the provision of support services to individuals with flexible working arrangements.

At this operational level, the 'real' time information requirements of FM will certainly warrant a different and more active management approach than paper-based or centrally controlled computer activities permit. The 'time' critical 'electronic' business environment, could help to build a co-ordinated service operations management approach of more relevancy to the customer. But there is a threat. Developments of this kind could conceivably replace whole areas of the FM function altogether. From the FM 'platform 2000' generally, the major uncertainties and dilemmas for the future include:-

- How to progress and develop the new forms of FM expertise that will be needed to support the dynamic and diverse knowledge based businesses and social needs of this new century?
- How to access the FM knowledge base and develop interactive tools to realise its value through viable vehicles for delivery and practical application?
- How to reconcile, interpret and adjust the 'generic' nature of the FM knowledge supply with the highly 'situation specific' character of FM knowledge demands in practice?
- How to determine which parts of the FM knowledge base could be exploited responsibly, effectively and reliably via the Web and other ICT technology, what not?
- How to decide which types of procurement, which types of FM services, to which types of customers, might be delivered reliably and securely through 'e' business, which not?
  - How to develop a professional framework and ethical approach that recognises and supports the interests of all parties and all sectors of the facility management service in the future?

This last dilemma is considered in Speculation 9 below. Bob Grimshaw looks at professional developments in FM, returning to the question of professional 'ethics', as raised earlier. He suggests that FM needs to champion end user needs and well-being in the workplace, alongside and balanced with the commercial and organisational support requirements. A recognition of the FM responsibilities to all stakeholders will be essential, to ensure that the contractual relationship between the facilities manager and employer does not dominate the FM function unduly. The concluding section of this final chapter considers this need for a broadly based professional framework, that serves the interests of all, along all of the FM trails to the future.

Speculation 9   **Facilities Management and Ethics**

Bob Grimshaw, University of the West of England

In 1994 the British Institute of Facilities Management (BIFM) decided that its future lay in becoming a 'professional' institution that would act as the gatekeeper for the profession of FM. It set in place three routes to a rigorous qualification and the infrastructure to operate and monitor these routes. All are now functioning and the first professionally qualified facilities managers have emerged. This strategy is long term and its success depends upon FM becoming widely recognised as a professional function.

Implicit in the view that FM is a profession (certainly as it is understood in the UK) and as opposed to the view of FM as a specialised field of general management, is a presumption that FM has a wider social importance outside of the contractual relationship between FM and employer. It implies that qualified facilities managers should operate within an ethical framework that influences and shapes the way they carry out their role.

In the established 'elite' professions, like law, medicine and architecture, their social relevance and ethical battlegrounds are clear: debate is on-going, vigorous and often controversial. No parallel debate has taken place in FM nor is there any established view on what the nature of an ethical stance for FM might be. Indeed, as FM develops rapidly towards becoming an 'organisational support mechanism', the grounds for an ethical debate are far from obvious. If FM cannot generate this debate and demonstrate its wider importance to society its claim to be a 'profession' looks hollow.

It is interesting to speculate on what the nature of FM ethics might be, over and above the obvious need to act in a professional manner in all one's dealings. One area is easy to identify but problematical, because FM appears to be moving away from it. The issue of how, or if, organisations should take account of user/employee needs in the workplace was prominent in early texts on FM. Frank Becker in particular saw FM as a channel of communication between the organisation and its staff in the commercial workplace. This was not justified in social terms but by sound economic arguments related to flexibility in a rapidly changing world. It could be argued that this even more relevant today as knowledge replaces product or service as the focus of many large organisations.

It is not too great a leap from Becker's work, supported by evidence from research on organisational change and behavioural science, to present FM as a profession that promotes humane working environments. Such environments, developed to take account of both work tasks and individual or group needs, have an obvious social relevance. The elements for and against such a position already exist but debate needs to be instigated by the BIFM and joined by practitioners and academics alike. The outcomes may be unpredictable but it would be healthy for a new profession to have a sense of its social value and a mission with which it enthuse its members. The alternative, which appears to be a sterile branch of management consultancy that can at best be called a 'quasi-profession' cannot be contemplated.

The scope of facility management practice continues to expand rapidly. It currently operates at three main levels. Firstly, FM is about operating buildings, the management of their facilities and support services at a single location. This is the domain of the facilities manager. It is the primary focus of concern for the members of the young 'professional' FM institutes such as the International Facilities Management Association (IFMA), established in 1980, and the British Institute of Facilities Management (BIFM), established in 1994. Equivalent organisations in other European countries and in Japan and Australia have a similar 'operational' focus.

Secondly, FM is directed to the strategic management of property portfolios and their facilities spread over a number of locations, with integrated services support. This is the domain of corporate real estate managers, facility directors and major facilities management companies. It is a natural focus for members of the established professions, such as the Royal Institute of Chartered Surveyors (RICS) and its Facility Management Faculty.

Thirdly, FM has a role in the global management of infrastructure, property, facilities and services to meet the demands of international business. This is the domain of the senior property and facility management consultants. It focuses on the strategic and policy levels of FM, usually involving senior professionals with MBA, MSc, or equivalent qualifications in the business and facility fields.

So FM currently has an extremely broad scope, some would say ridiculously so. It serves a wide variety of sectors and types of organisation, with quite different cultures, objectives and modes of operation. Its management roles are also wide, with local, operational, strategic and global levels of application. The principal dilemma for facility management from its 'platform 2000', is a result of its own success and this extensive and expanding scope. But what will be the future scope of FM be? Will it stabilise? And given this ever widening scope, how should FM organisation its own future. How should those working in the field best organise themselves professionally?

It is most unlikely that the broad cross-disciplinary territory that is covered by FM could ever be ring-fenced by a single professional body, claiming it as its own. However some do suggest that facilities management has emerged as a new profession in its own right. For them FM is seen as a clearly defined and inclusive area of professional activity, responsibility, competency and accountability. Others see the field of FM fragmenting into a number of specialist fields, both within general business administration and management on the one hand, and as additional areas of competency and responsibility within the traditional property professions, on the other hand.

These two views are not mutually exclusive however. Again there is a 'third way'. FM could become a truly multi-professional undertaking. Progress here, would be reliant on the development of an 'open' cross-disciplinary and multi-profession culture to provide internal coherence for those operating across the diversities of the field, and external coherence to the business community, the property professions, government agencies and the public.

Within this 'multi-professional' model, the British Institute of Facilities Management would be likely to continue to play its crucially important role as the national focus for 'operational' FM. But to move forward, BIFM should avoid the attitudes of 19th century professionalism and distance itself from the 'closed' and 'inclusive' self-interests of the older professional bodies. Success at this local level will depend on BIFM's implementation of its challenging programme of measures to improve facilities managers' professional skills, competencies and accountability in operating buildings, facilities and support services.

Similar institutions in other countries would do likewise, but with unavoidable and desirable differences in FM practice dependent on the particular national culture, social and economic circumstances, climate, laws and regulations of their region. The International Facility Management Association (IFMA), working out of its North American base, will hopefully continue to provide an invaluable international perspective of professional operations in FM through 'cross-national' and 'cross-sector' benchmarking But it needs to demonstrate that it is not intent on establishing a form of 'global imperialism' in FM. All should take care to avoiding the fallacy of the 'universal solution' and other tendencies towards uniformity.

So the range of national and international FM institutes would remain at the operational centre of FM in a 'multi-professional' future. They would need to interface with the very wide range of professional organisations that have legitimate interests and advanced expertise in specific areas of the FM domain. Examples of areas requiring active collaboration in the UK context will include; facility strategy and finance with the RICS (The Royal Institute of Chartered Surveyors), environmental management, building services and energy management with CIBSE (The Chartered Institute of Building Services Engineers), the support of human resources with IPD (The Chartered Institute of Personnel and Development), business infrastructure management with the leading business schools and the CBI (The Confederation of British Industry), facility briefing, design and adaptation with the RIBA (The Royal Institute of Architects), project and construction management with the CIOB (The Chartered Institute of Builders), procurement management with the CIPS (The Chartered Institute of Purchasing and Supply) and there are many more.

But this will only be a beginning. The 'big' challenge for FM will be to build a 'multi-professional framework for applied research, development and application, across this wide range of professional activity, in collaboration with the relevant government agencies. Research organisations and government agencies will only become involved in FM initiatives if they are convinced that the proposed work is relevant to their policies and programmes, and that it promises, in some small way, to contribute to the national interest. The challenge for the FM is to demonstrate how research and development might result in improved infrastructure, facilities and services that provide better strategic and logistic support to the nation's social and business endeavours of all kinds.

The key areas for development will lie between, rather than within, the boundaries of the traditional professions. New knowledge links must therefore to be forged between those in management and those in design, between the property professions and business, between those in HR management and those with expertise in the management of business operations, facility logistics and services support. These intersections suggest 'professional diversity' as the key to future success. Progress in FM will be reliant on diversity, variety and versatility, building on an open cross-disciplinary and multi-professional approach that is applicable and adaptable to all sectors. A single integrated future for facility management is neither viable or desirable.

Managing this diversity will be the key to the future.

References

The Economist (2000) Internet Economics, 2nd April, pp. 77-79.

Kincaid D. (1994) Integrated Facility Management, *Facilities,* Vol. 12, No. 8, pp. 20-23.

Nutt B. (1991) Agenda for Education, *Facilities,* Vol. 9, No. 5, pp. 9-14.

Nutt B. (1998) Collaborative Research, *Facilities Management World*, Issue 8. Jan/Feb, pp. 11-13.

Thackara J. (1999) Conference Transcript from "Linking Management and Design Knowledge", *Conference: Futures in Property and Facility Management,* UCL, London.

Tickell C. (1999) Conference Transcript from New Strategic Directions, *Conference: Futures in Property and Facility Management*, UCL, London.

Index